National Défense
Defence nationale

**B-GA-007-001/PT-D01**

METEOROLOGY

# AIR COMMAND WEATHER MANUAL

(ENGLISH)

**(Supersedes B-GA-007-001/PT-D01 dated 2004-12-17 also known as CFACM 2-700 dated 1987-10-30)**

Issued on Authority of the Chief of the Defence Staff and the Deputy Minister of National Defence
Publié avec l'autorisation du Chef d'état-major de la Défense et du Sous-ministre de la Défense nationale

OPI: 1 Cdn Air Div, A3 Met

2008-05-30

**1 Canadian Air Division**

Published for 1 Canadian Air Division by the 17 Wing Publishing Office.

Printed in Canada.

**Library and Archives Canada Cataloguing in Publication**

Main entry under title :

Air command weather manual

"B-GA-007-001/PT-D01"
"Supersedes B-GA-007-00/PT-D01 Change 1 dated 1987-02-18 also known
as CFACM 2-700 dated 1987-10-30"
Issued also in French under title: Manuel de météorologie du commandement aérien.
"The Air Command Weather Manual is the official reference for meteorology for aircrew in the
Canadian Forces and has been adopted by Transport Canada for civilian use."—Pref.
Companion volume to the Air command weather manual workbook.
ISBN 0-660-19348-5
Cat. No. D2-161/1-2004E

1. Meteorology in aeronautics – Canada – Handbooks, manuals, etc.
2. Meteorology in aeronautics – Handbooks, manuals, etc.
3. Navigation (Aeronautics) – Canada – Handbooks, manuals, etc.
I. Canada. Dept. of National Defence.
II. Canada. Canadian Armed Forces. Canadian Air Division, 1.

TL558.C3A37 2004          629.132'4          C2004-980273-9

# TABLE OF CONTENTS

**Table of Contents continued...**

**Table of Contents continued...**

**Table of Contents continued...**

**Table of Contents continued...**

**Table of Contents continued...**

**Table of Contents continued...**

B-GA-007-001/PT-D01

**Table of Contents continued...**

# LIST OF FIGURES

**List of Figures continued...**

B-GA-007-001/PT-D01

**List of Figures continued...**

**List of Figures continued...**

**List of Figures continued...**

# PREFACE

1.  The Air Command Weather Manual is the official reference for meteorology for aircrew in the Canadian Forces and has been adopted by Transport Canada for civilian use.

2.  This manual provides an introduction to weather fundamentals and more in-depth information on aviation weather in particular.

3.  Knowledge of the sky, while essential, is not enough. An aviator must be able to apply this knowledge to actual flying. Through the practical application of learned material you can develop the skill and judgement essential to your success as an aviator.

# FOREWORD

1.  The Air Command Weather Manual is issued on the authority of the Chief of the Defence Staff and the Deputy Minister of National Defence.

2.  The is effective upon receipt and supersedes CFACM 2-700 dated 1 October 1984.

3.  Suggestions for amendments should be forwarded through normal channels to 1 Canadian Air Division Headquarters, Attention: A3 Meteorology.

30 May 2008

# Chapter

# 1

You are going to have to make decisions regarding the weather many times in your flying career. Since most weather forms as water changes between vapour, liquid and ice, some knowledge of the behaviour of moisture in the atmosphere will help you make the correct decisions.

# CHAPTER 1

## MOISTURE IN THE ATMOSPHERE

1. Clouds and precipitation are obvious forms of water in the atmosphere, but there is water vapour mixed with the air even on a bright, cloudless day. The amount of the water vapour varies and, although we cannot sense it as easily as air temperature, we can feel the difference in the humidity of a hot, muggy day as compared to that of a cold, crisp one.

2. A very important exchange of energy occurs when water in the atmosphere changes between vapour and cloud which is basic to the development of weather. On the scale of atmospheric processes, the amount of energy in this change is far from trivial. The energy associated with one thunderstorm, for instance, comes mainly from this source and is equivalent to a dozen or so Hiroshima-type bombs. A hurricane releases almost that much energy in a second.

### Changes of State

3. The moisture in the atmosphere originates mainly from evaporation from the oceans and lakes and from transpiration from the vegetation of the earth's surface. While the amount remains greatest near these sources it is mixed throughout the lower 20,000 to 40,000 feet of the atmosphere. Since cloud is formed when water vapour changes to liquid water droplets or ice crystals, it is only within this layer that clouds and precipitation occur.

**Figure 1-1 Source of atmospheric moisture**

4.  Water can change back and forth from gas, liquid and ice at ordinary atmospheric pressures and temperatures. As a gas, water is in a high energy state, with its molecules moving freely and rapidly. As a liquid it is in a medium energy state and as ice it is in a low energy state with the molecules moving only slightly. Water as high energy water vapour can condense to lower energy liquid and remain at the same temperature. As it slips into this lower energy state it releases energy to the atmosphere as heat. This heat is called the "Latent Heat of Vaporization." In the reverse process when liquid evaporates to vapour or gas at the same temperature, it absorbs the same amount of energy from the atmosphere. This is what causes cooling of the skin from perspiration. In this case, the heat required to evaporate the perspiration is taken from the skin, thus cooling it.

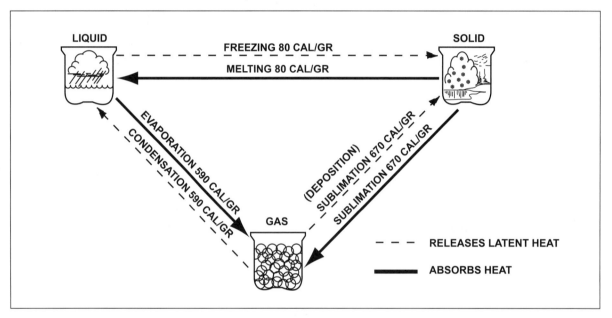

**Figure 1-2 Change of state**

5.  During the change from liquid to ice, heat is released into the atmosphere, and from ice to liquid, heat is absorbed. In these cases the heat gained or lost is called the "Latent Heat of Fusion." Ice can change directly to water vapour and water vapour directly to ice. In both cases the process is called "Sublimation" and the heat gained or lost is called the "Latent Heat of Sublimation." Heat is released in the change to ice and absorbed in the change to vapour and it is equal to the latent heat of vaporization plus the latent heat of fusion. To avoid confusion, the term "deposition" is starting to be used when describing the change of state of gas directly to ice. The energy previously referred to in thunderstorms or hurricanes comes from the release of the various latent heats as water vapour changes to water droplets and then to ice crystals.

## Moisture Content

6.  There is a limit to the amount of water that can exist as vapour in the air at any given temperature. When this limit is reached, saturation occurs and any cooling will cause condensation and cloud will form. This is illustrated in Figure 1-3, which shows the water vapour content when the air is saturated at various temperatures. From this diagram you can see that at 30°C, 30 grams of water vapour can

exist in a cubic metre of air (point A). At 27°C, only 25 grams per cubic metre can exist (point B), at 6°C, 7 grams (point C) and at 3°C, 6 grams (point D). If the temperature drops 3°C from 30°C to 27°C, 5 grams of water vapour per cubic metre will condense. A similar 3°C drop from 6°C will cause only 1 gram of water vapour per metre to condense. Cooling of saturated air at warmer temperatures causes more water vapour to condense than the same amount of cooling of saturated air at colder temperatures.

7. Relating this fact to latent heat and energy, more energy is released during condensation at warmer temperatures than at cooler temperatures for an equivalent amount of cooling. For this reason, the most violent weather, such as hurricanes, tornadoes, and severe thunderstorms, occurs in very warm, moist air.

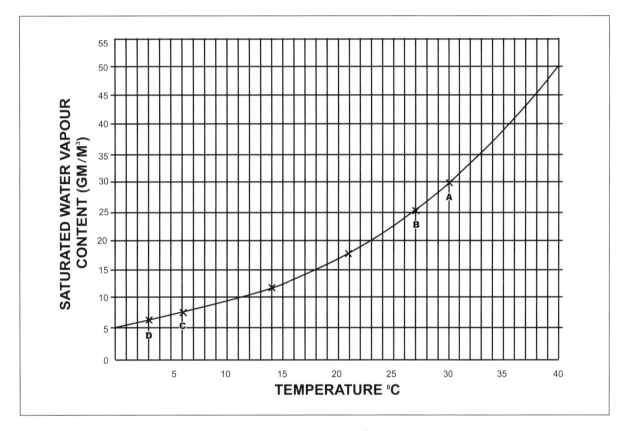

**Figure 1-3 Variation of saturated water vapour content with temperature**

## Dew Point

8. You can assess the amount of water vapour in the atmosphere from the dew point. The "Dew Point" is the temperature at a given pressure to which air must be cooled to cause saturation. The difference between the temperature and the dew point is an indication of how close the air is to saturation. This difference is referred to as the "Spread," and a spread of zero means that the air is saturated. The higher the dew point, the more water vapour there is in the air. Look at Figure 1-3. You can see that at a dew point of 30°C there are 30 grams of water vapour in a cubic metre of air, but at 6°C there are only about 7 grams.

## Relative Humidity

9.  The "Relative Humidity" compares the amount of water vapour in the air with the amount that it could hold if it were saturated and this is expressed as a percentage. One hundred per cent relative humidity indicates saturation. Look again at Figure 1-3. Air with a 100% relative humidity at 30°C holds 30 grams of water vapour per cubic metre but at 6°C it holds only 7 grams. In both cases, there is 100% relative humidity, but the actual water content is quite different. Relative humidity, therefore, does not give as direct an indication of moisture as does dew point. For this reason, dew point and not relative humidity is provided for most aviation purposes.

## Condensation Nuclei

10. In the condensation process, water changes from its gaseous state into tiny water droplets, and in the sublimation process, it changes to tiny ice crystals. A peculiar feature of these processes is that they will not occur even when the air is cooled below the saturation temperature unless small particles called condensation or sublimation nuclei are present in the atmosphere. The nuclei are formed of sea salts from the evaporation of ocean spray or of tiny solid particles formed during combustion in industrial areas or from forest fires. They are always present in sufficient numbers that condensation will occur at 100% relative humidity. If they are present in very large numbers, however, for example in industrial areas or over oceans, they can cause condensation to occur even before the air is completely saturated. This is one reason why there is so much cloud over the oceans and industrialized areas.

## Aircraft Performance

11. The density of air decreases with decreased atmospheric pressure, increased temperature and increased water vapour content. Aircraft performance varies with the air density so that at any given pressure the performance is better in denser air. Although temperature is the major controlling factor in this situation, moisture does play a small part so that for some critical aircraft operations, the moisture content of the air must be considered.

## Summary - Chapter 1

- Cooling causes saturation; further cooling causes condensation or sublimation and the formation of cloud.

- The amount of water condensed is greater when warmer saturated air is cooled than when colder saturated air is cooled.

- During evaporation, melting and sublimation (ice to vapour), heat is absorbed by the moisture.

- During condensation, freezing and sublimation (vapour to ice), heat is released to the atmosphere.

- The dew point is the temperature to which air must be cooled (at constant pressure) to cause saturation.

- The relative humidity is the amount of water vapour in the air compared with what it would hold if it were saturated.

- Nuclei must be present for condensation or sublimation to occur. If they are present in abundance, these processes can occur at less than 100% relative humidity.

- Aircraft performance decreases slightly in air having a high water vapour content.

# Chapter

## 2

Ultimately, solar energy is the driving force causing all the changes and weather patterns in the atmosphere that affect flying.

# CHAPTER 2

## THE ATMOSPHERE AND HOW IT IS HEATED

2 CHAPTER

### Introduction

1. The atmosphere is composed of a mixture of gases, each invisible and each adding its percentage to the total volume of the air surrounding the earth. Although water vapour is only a small percentage of the total volume of gases in the atmosphere, it is very important from the standpoint of weather. Water can exist as a liquid, a gas or a solid under atmospheric conditions; thus when water in the gaseous state changes to water in the liquid or solid state, ie, water droplets or ice crystals, it forms either cloud or fog, depending on the altitude at which the change takes place. One of the major problems that confronts the forecaster is the determination of exactly when and where water vapour will change into a visible form. The difficulties associated with forecasting cloud and fog formation are increased by the fact that, unlike other gases in the atmosphere, water vapour varies in amount from day to day and even from hour to hour. The air is said to be dry when there is no water vapour mixed with it; when the air has water vapour contained in it, it is said to be moist.

### Composition

2. The atmosphere contains many important gases in greater or lesser concentrations. Dry air is composed of 78% nitrogen, 21% oxygen and 1% argon. Smaller portions of helium, neon, ozone and carbon dioxide are also found in gases that surround the earth.

3. In addition to the gas in the atmosphere, the lower levels contain quantities of solid particles that are important to both forecasters and aircrew. These particles can reduce the visibility through the air, and are also important in the process of condensation (gas to a liquid) and sublimation (gas to a solid). If no solid particles were in the air, it would be very difficult for cloud droplets to form and thus for weather to occur.

4. The solid particles in the atmosphere are varied in size, shape and composition. In size, they range from submicroscopic to large elements that can be detected by the eye. They may be organic such as seeds, pollen, spores or bacteria or they may also be inorganic elements such as soil, smoke or salt from ocean spray. These solid particles, or cloud condensation nuclei, are more concentrated near the surface of the earth than they are at a high altitude. This is evident when one flies at a high level; the visibility is normally much better at 40,000 feet than it is at 1,000 feet above ground. During high wind conditions, particles can be carried to high levels with time; however, the particles fall back to the ground.

5. Concentrations of these solid particles change with the location. The number of particles per unit volume is much higher near cities and industrial areas than in rural areas or over the sea.

CHAPTER **2**

## Properties of the Atmosphere

6.   The principal properties of the atmosphere are mobility, the capacity for expansion and the capacity for compression. Although the atmosphere is often referred to as an ocean of air and the winds are compared to streams of water, it is necessary to remember that there is a great deal more freedom of motion in the air than there is in the water. In other words, the greater mobility of the air must be kept in mind.

7.   A given parcel of air is capable of indefinite expansion, for the enclosed gases themselves exert a pressure tending to change their volumes. If the pressure surrounding the parcel is decreased, the gases within push the parcel walls further apart, resulting in an internal decrease in temperature. This property of expansion is of the utmost importance in the study of weather, as it is the reason for much of the cloud and the weather that we see from day to day. Portions of the air in the atmosphere are often forced to rise due to heating from below or winds forcing air up a slope. As the air rises, it reaches regions of lower pressure, expands and cools. In many cases the cooling is sufficient to cause condensation of the water vapour in the air. It is for this reason that clouds and precipitation are common in areas of rising air.

8.   Associated with the capacity for expansion is the ability of the air to be compressed. If air is subjected to an increase in external pressure, the volume decreases and the temperature of the air increases. Under natural conditions, large blocks of air are forced to descend to where the pressure is higher. The resulting compression causes the volume of the air to decrease and the temperature of the air to increase. The resulting increase in the temperature of the air causes the water droplets, if there are any, to evaporate. It is for this reason that areas of descending air tend to be areas which are cloud-free.

## Divisions and Characteristics

9.   In meteorology, the atmosphere is divided into different layers according to the characteristic of the temperature in that layer, as indicated in Figure 2-1. The atmosphere is divided into four distinct layers depending on the temperature and the changes of temperature with height. The layers of the atmosphere are, in ascending order, the troposphere, the stratosphere, the mesosphere and the thermosphere. The top of each layer is known as the tropopause, stratopause and the mesopause respectively.

10.   The "troposphere," the layer nearest to the earth, is normally identified by a decrease in temperature throughout its depth. The point at the top of the troposphere at which the decrease stops and the temperature starts to increase is known as the "tropopause." The average height of the tropopause is about 11 kilometres. The height of the tropopause can vary considerably, however, and be in the order of 17 kilometres over the equator and about 8 kilometres over the poles. The tropopause is generally higher in the summer than in the winter.

11.   Most of what we call "weather" occurs in the troposphere. This is mainly because of the presence of water vapour and large-scale vertical currents in this part of the earth's atmosphere. In its cold upper regions near the tropopause, winds reach maximum speeds and become complex in structure, owing to the presence of narrow, rapidly moving streams of air known as "jet streams." These jet streams are embedded in the general flow and can reach speeds of 200 knots or more. Near a jet stream, the height of the tropopause can change abruptly in only a few miles.

12. The region above the tropopause is known as the "stratosphere." This layer is the region of the earth's atmosphere where the temperature of the air remains the same with height and then increases up to the top. The steady increase of temperature to the top of the stratosphere is due to minute amounts of ozone absorbing ultraviolet rays from the sun. This causes the temperature of the air to rise to near zero at the top of the stratosphere. The thickness of the stratosphere varies considerably; it is much thicker over the poles than over the equator where it may be non-existent. Clouds in the stratosphere, although very rare, still occur on occasion. These clouds, known as "nacreous clouds," or mother-of-pearl, form at high levels and are thought to consist of ice crystals. The point at the top of the stratosphere where the temperature of the air again starts to decrease is known as the "stratopause." Aircraft flying in the stratosphere can look forward to excellent visibility, no turbulence and no cloud at flight levels currently used.

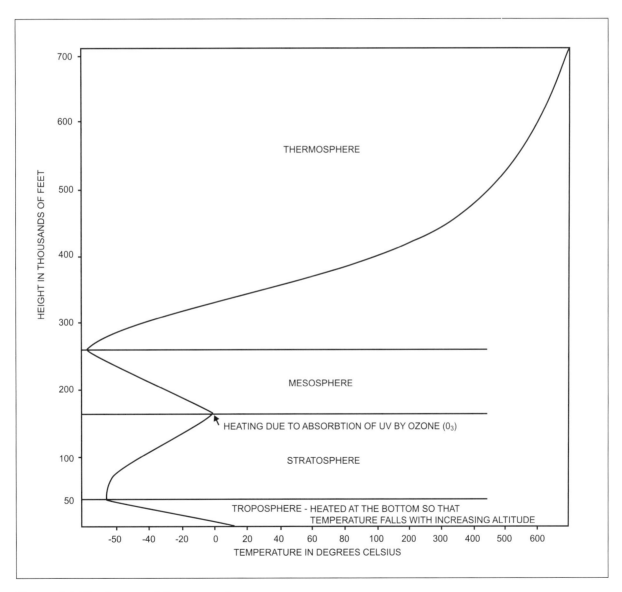

**Figure 2-1 The layers of the atmosphere**

13. The "mesosphere" is the layer of the atmosphere found immediately above the stratosphere. This layer is characterized by a decrease in temperature to about the 275,000 foot level. The point of lowest temperature is known as the "mesopause." Above the mesopause, the temperature again increases dramatically to values in the thousands of degrees Celsius. This layer, where the temperature increases, is known as the "thermosphere." Because of the thinness of the air, the temperature of this air would not be felt on the human body and is only a kinetic temperature, which governs the speed of the molecules in the thermosphere. The aurora is found in this layer of the atmosphere and is caused by particles from the sun causing molecules of oxygen, hydrogen and nitrogen to fluoresce.

14. Overall aircraft efficiency improves with cold temperatures. Since atmospheric temperatures are generally coldest at and just above the tropopause, its height is of considerable importance. While the average height of the tropopause is near 36,000 feet, its actual height constantly varies and is primarily dependent on the temperature of the troposphere. The warmer the troposphere, the higher the tropopause and, conversely, the colder the troposphere, the lower the tropopause. The warmth of the troposphere is dependent on the amount of surface conduction and terrestrial radiation that it receives.

15. In Figure 2-5, the same intensity of solar radiation strikes the earth's surface in each case; however, the beam of radiation striking at an angle is spread over a larger area so the radiant energy received per unit area is less. For this reason, the troposphere will tend to be warmer where the solar rays are perpendicular to the earth.

## Radiation

16. All matter radiates energy in the form of electromagnetic waves. The amount of energy and the wavelengths emitted are dependent on the temperature of the matter. The hotter a substance is, the greater the amount of energy emitted and the shorter the wavelength of the emission.

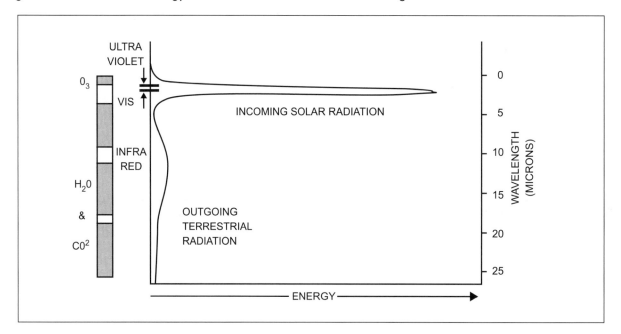

**Figure 2-2 Solar and terrestrial radiation**

17. Figure 2-2 shows the radiations emitted by the sun and by the earth. The major portion of solar radiation is from ultraviolet, through normal light visible to the eye, into the infrared. Radiation from the earth lies in the infrared range and blends from the solar wavelengths into longer infrared wavelengths.

18. Electromagnetic radiation is not heat; however, if the radiation should be absorbed by a substance or a gas it will heat the substance or gas and cause a rise in temperature. Gases are selective in the wavelengths that they absorb and permit certain bands of wavelengths to pass through unhindered. These wavelengths are called "Windows." Other bands of wavelengths are absorbed and cause the temperature of the gas to rise. The left hand portion of Figure 2-2 illustrates the absorption bands and windows. The shaded areas are opposite wavelengths that are absorbed, the clear areas are opposite windows. Terrestrial radiation is absorbed in the lower few thousand feet, primarily by the water vapour and carbon dioxide which are constituents of the lower atmosphere. In the upper atmosphere, ozone absorbs some of the solar radiation. Other gases in the atmosphere are not significant in absorbing radiation.

## Absorption of Solar Radiation in the Ozonophere

19. Ozone is concentrated in a layer called the "Ozonosphere" extending from about 33,000 feet up to 165,000 feet. It absorbs most of the solar ultraviolet radiation. If this ultraviolet radiation were to reach the earth's surface it would kill all life on earth. Ozone, however, is not totally beneficial since it can cause sickness in human beings and can corrode metals. For these reasons, precautions must be taken against ozone while flying within the ozonosphere.

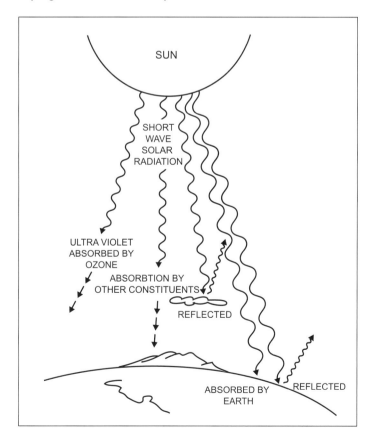

**Figure 2-3 Solar radiation**

20. The absorption of solar ultraviolet radiation by ozone raises the temperature of the high atmosphere to near the freezing point. As the radiation penetrates further into the ozonosphere, the ultraviolet radiation is gradually depleted by absorption and the temperature of the atmosphere steadily falls to a minimum at a height that averages around 36,000 feet. The remainder of the solar radiation continues through the atmosphere with only a small amount of it being absorbed by other atmospheric constituents. Some of it is reflected back to space from the atmosphere, cloud tops or the earth's surface. The remainder is absorbed by the earth's surface, which is warmed, and reradiates in longer, infrared wavelengths. (Figure 2-3)

## Absorption of Terrestrial Radiation

21. Some of the outgoing infrared radiation from the earth's surface is absorbed by carbon dioxide and water vapour in the atmosphere, as shown in the left hand side of Figure 2-2. The amount of heating that this creates decreases with altitude as the radiation is depleted. Cloud, if present, absorbs a great deal of terrestrial radiation and reradiates it both back to earth and out to space. Some of the terrestrial radiation passes directly out to space through windows, and the atmosphere itself radiates out to space. The outgoing radiation balances the incoming radiation from the sun so that the earth's average temperature remains nearly constant. (Figure 2-4)

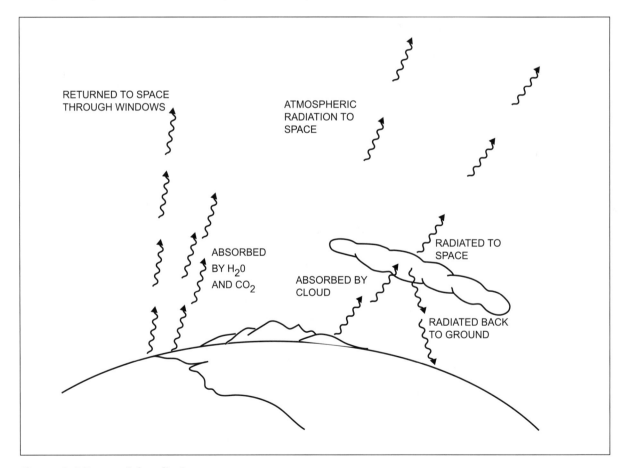

**Figure 2-4 Terrestrial radiation**

## Conduction

22. A law of physics states that if two bodies are touching, heat will flow from the warmer to the colder body. A layer of air touching the earth's surface that is warmer than itself will be heated by conduction. Air is a very poor conductor so the heat received in this way will remain within a very shallow layer unless it is distributed aloft through vertical air motion.

23. The combined effect of terrestrial radiation and conduction causes the lower several thousand feet of the atmosphere to be heated from below.

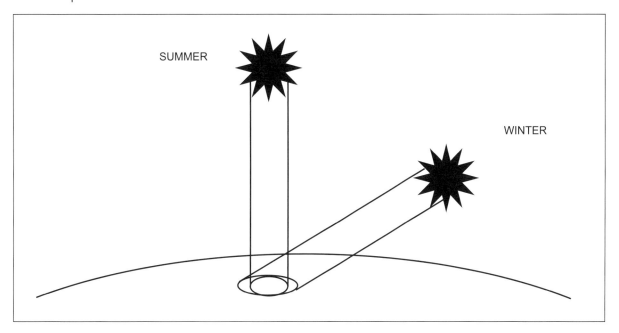

**Figure 2-5 Angle of incidence**

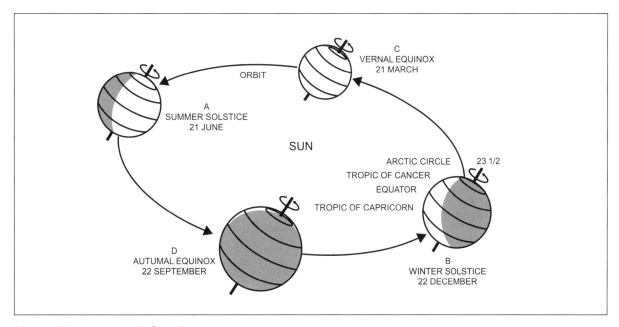

**Figure 2-6 Astronomical setting**

## Astronomical Setting

24. The earth rotates about its axis once in approximately 24 hours and orbits the sun in a little over 365 days (Figure 2-6). The earth's axis is tilted so that the vertical rays from the sun giving maximum insolation swing back and forth across the equator as shown in Figure 2-7. This produces summer and winter in each of the hemispheres with the tropopause rising or falling as the troposphere heats or cools.

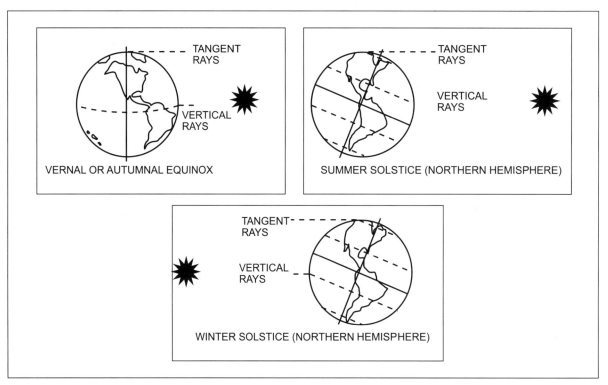

**Figure 2-7 Belts of maximum insolation**

## Reflection

25. Some of the solar energy that strikes the earth's surface is reflected back out to space and does not heat the surface. The amount of reflection depends on the angle at which the radiation strikes the surface and on the type of surface it strikes. A snow surface, for example, can reflect 90% of the radiation. On a bright, sunny winter day on the Canadian Prairies, the air temperature may stay at -30°C throughout the day simply because the solar radiation is reflected from the surface snow and is not absorbed. In general, surfaces that are good reflectors will absorb only small amounts of solar radiation and the solar rays that strike the earth at a shallow angle are largely reflected.

## Maritime Effects

26. The surface temperature of lakes or oceans does not change as much or as rapidly as land surfaces. It takes about five times as much radiant energy to raise the temperature of water as it does to raise the temperature of dry earth. Heat received at the water's surface can be distributed downward to a considerable depth by water motion, whereas heat received by soil is held within a few inches of the surfaces.

27. Because of these characteristics, there are fundamental differences between land areas and water areas on the earth's surface:

    a. With the same amount of solar energy falling on each surface, a land surface will reach a higher temperature more quickly than a water surface. Also, when the supply of solar energy is removed during night-time, a land surface will cool more rapidly.

    b. Land or continental areas will be characterized by large diurnal (day to night) and seasonal temperature ranges.

    c. Water or maritime areas will be characterized by very small diurnal and seasonal temperature ranges.

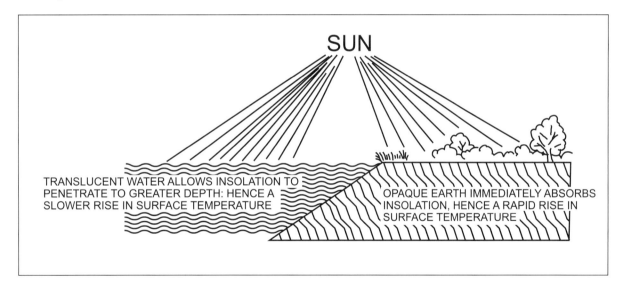

**Figure 2-8 Heating of land and water surfaces.**

## How the Troposphere is Heated

28. The troposphere receives its heat from terrestrial radiation and from conduction. The heat is distributed throughout the troposphere by several other processes. These are convection, turbulent mixing, the release of latent heat and advection.

## Convection

29. The surface layer of air heated by conduction becomes buoyant and rises up through the atmosphere as a convective current that carries surface heat upward into the atmosphere.

## Turbulent Mixing

30. Wind causes turbulent air motion that mixes the surface layer of air that has been heated by conduction with the unheated air aloft, thus spreading the heat upwards.

## Latent Heat

31. Water vapour evaporated into the atmosphere from the earth's surface is frequently carried aloft. If it should condense, its latent heat is released. This heat, which originated near the earth's surface in the evaporation process, is thereby distributed in higher levels of the atmosphere by condensation.

## Advective Warming

32. Air being carried from a cold portion of the earth's surface to a warmer portion by wind will have its lowest levels heated by conduction and the heat will be distributed upward by convection and turbulent mixing. The term used to describe warming in this manner is called "Advective Warming."

## Compression

33. There are occasions when large sections of the earth's atmosphere subside. This would occur in the instance of air flowing down the side of a mountain range. As the air descends, it comes under increased atmospheric pressure and is compressed. This compression heats the subsiding air. A commonplace example of compression heating is the heat produced in a hand-pump when pumping up a bicycle tire.

## Temperature

34. For aviation purposes in Canada, air temperature is provided in degrees-Celsius (°C). The surface temperature is measured in a ventilated louvered box at about four feet above the ground to eliminate any radiation effects. On some weather charts, lines are drawn joining places having the same temperature. These lines are called "Isotherms."

## Temperature and Aviation

35. Although air temperature is indirectly important to aviation because it is related to the development of weather, it is also directly important for several reasons. For instance, thrust, drag, lift, heating and cooling requirements are all affected by the air temperature so that all the possible variations in air temperature that an aircraft may encounter must be considered when it is being designed.

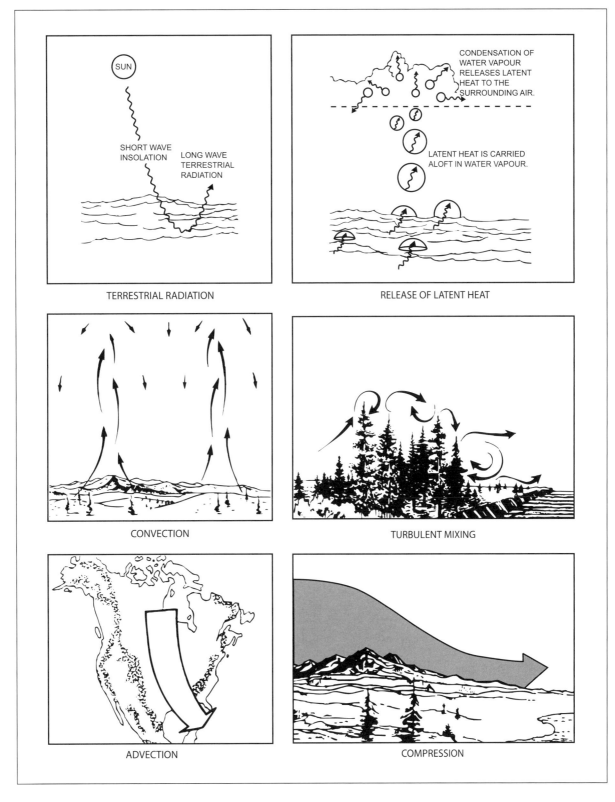

**Figure 2-9 How the troposphere is heated**

36. The performance of an aircraft depends on several factors, among which temperature is important. The efficiency of a jet engine depends in part on the difference between the outside air temperature and the maximum temperature attainable in the combustion chamber. When the air temperature increases above a certain value, depending on the altitude, the true airspeed and the aircraft efficiency both fall off, the aircraft's operating height is reduced and there is an increase in fuel consumption per mile. Take-off performance too, is strongly affected, being much worse at very high temperatures.

## Mach Number

37. During flight, an aircraft sets up pressure waves that spread out at the speed of sound, in all directions. If the aircraft approaches the speed of sound, these waves are compressed ahead of it. This disturbs the flow pattern over the aircraft and has an adverse affect on the lift, drag and stability of the aircraft and its reaction to the flight controls. These problems become noticeable at about two thirds the speed of sound and become progressively worse as speed increases up to the speed of sound but improve again beyond the speed of sound.

38. The speed of sound is not constant but varies with the air temperature. At 15°C it is about 660 knots and at -56°C, 575 knots. In order to indicate what the airspeed is in relation to the speed of sound a "Mach number" is provided. It is simply the ratio of the true airspeed to the speed of sound under the air temperature conditions that the aircraft is flying in. A Mach number of .70 means that the aircraft is flying at 7/10 of the speed of sound for the air temperature being encountered. Mach 2.0 indicates that the aircraft is flying at twice the speed of sound. The Mach number is often indicated in the aircraft by an instrument call a "Machmeter."

# Summary - Chapter 2

- The atmosphere is composed of the troposphere, stratosphere, mesosphere and thermosphere.

- The stratosphere is heated by ozone absorbing solar ultraviolet radiation.

- The troposphere is heated from the earth's surface by absorption of terrestrial infrared radiation and by conduction.

- The atmospheric temperatures are coldest at the tropopause.

- The tropopause is highest over a warm troposphere and lowest over a cold troposphere.

- The surface temperature is dependent on the angle of the incidence of solar radiation, the amount of radiation reflected and the type of surface (land, water).

- Heat is distributed vertically throughout the troposphere by: convection turbulent mixing and the release of latent heat.

- The atmosphere can also be heated by compression and by advection over a warmer surface.

- Air temperature is provided in degrees Celsius.

- Isotherms are lines drawn on a weather chart which join places of equal temperature.

- Aircraft performance is reduced with high temperatures.

- The true airspeed for a given Mach number is higher in warmer air than in cooler air.

# Chapter

## 3

Cloud, fog and precipitation form when moist air is cooled. It will be important for you to recognize the flight situations where this cooling is likely to occur

# CHAPTER 3

# ATMOSPHERIC COOLING

1.  Atmospheric cooling is important to you as an aviator for two main reasons. The first of these is that it increases the density of the air. (Later you will learn how this creates some very unusual and sometimes hazardous low-level winds.) The other reason is that when moist air is cooled, cloud, fog and precipitation will form. Since cloud and fog at low levels are of major concern in flying, cooling in the lower levels of the atmosphere is particularly important.

## Radiation Cooling

2.  The troposphere is heated from the earth's surface by terrestrial radiation and conduction. As a result, the temperature normally decreases with altitude up to the tropopause and this decrease in temperature is called the "Lapse Rate." The temperature through the atmosphere at any particular time and place is called the "Environmental Lapse Rate" (ELR), and is frequently around 2°C/1,000 feet. (Figure 3-1).

3.  The earth becomes warmer, as long as it absorbs solar radiation. After the sun has set, the earth's surface continues to radiate and its temperature begins to drop (Figure 3-2). The atmosphere, on the other hand, does not cool appreciably by radiation after sunset since it is a poor radiator. The surface layer of air that is touching the earth, however, cools as the earth cools, due to conduction. Figure 3-3 illustrates how the environmental lapse rate changes as the surface temperature cools from that at 18:00 hours (in the afternoon) to that at 02:00 hours (in the night).

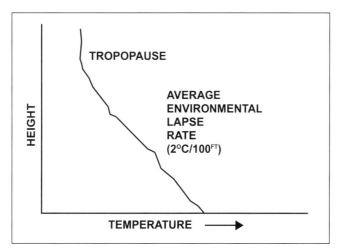

**Figure 3-1 Normal daytime environmental lapse rate**

4.  The radiation cooling that occurs at night will seldom affect more than the lower 4,000 feet of the atmosphere. The temperature above this will remain virtually unchanged from day to night. This does not apply to the long winter Arctic night however. In this case, the continual radiation cooling throughout the winter can cool the entire troposphere.

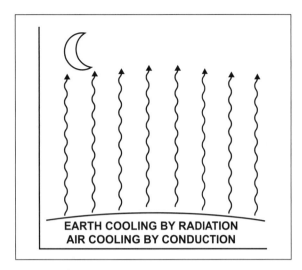

**Figure 3-2 Nocturnal radiation cooling**

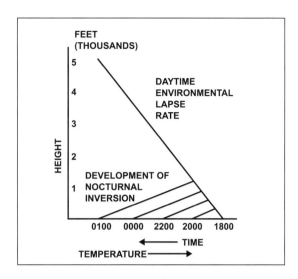

**Figure 3-3 Development of a nocturnal inversion**

## Nocturnal Inversions

5.   A temperature increase with altitude such as develops in Figure 3-3 is called an "Inversion." In this case, since the inversion has been caused by night-time cooling, it is called a "Nocturnal" or a "Radiation Inversion." The top of the inversion can frequently be seen on an early morning flight as the very sharp top of a surface-based haze layer. It will disappear as the sun heats the earth's surface. The temperature at the top of the inversion can be 15° to 20°C warmer than at the surface and is typically at around 1,000 feet above ground.

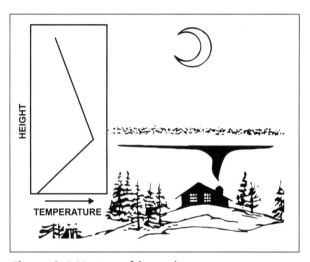

**Figure 3-4 Nocturnal inversion**

## Wind Effect

6.   On a windy night, turbulence mixes the lower few thousand feet of the atmosphere and distributes the cooling effect throughout this layer (Figure 3-5). Because of this, the temperature of the surface layer of atmosphere does not drop as much and the inversion is much weaker.

## Cloud Effect

7.   A blanket of cloud, particularly at low levels, absorbs terrestrial radiation and reradiates some of it back to earth (Figure 3-6). This also slows down the rate of cooling so that on a cloudy night the surface temperature does not drop as much as on a clear night.

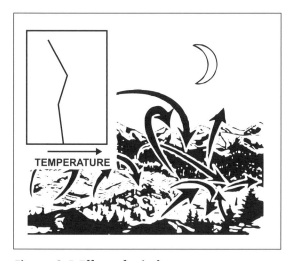

**Figure 3-5 Effect of wind**

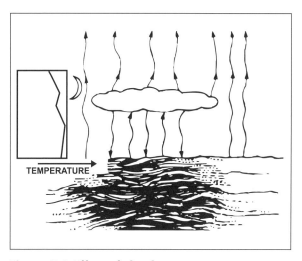

**Figure 3-6 Effect of clouds**

**Figure 3-7 Drainage effect**

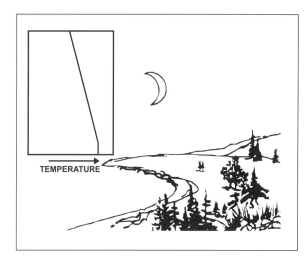

**Figure 3-8 Maritime effect**

## Topographical Effect

8.   Cold air is denser than warm air and at night it will flow into low-lying areas just as water flows downhill (Figure 3-7). Inversions formed where the cold air becomes trapped in low-lying areas, such as in valleys, can become very strong. On slopes, where the air can flow away, the inversion will be much weaker.

## Maritime Effect

9.   Much more heat energy is required to raise the temperature of a body of water than that of dry soil and this heat is distributed through a considerable depth of the water compared to an inch or two of the soil. When the sun has set, water continues to radiate just as does soil, but because there is so much more heat available in the water, its temperature drops only a very small amount. For this reason, in maritime areas, the nocturnal inversion is very weak or non-existent (Figure 3-8).

## Adiabatic Process

10. If air, for some reason, should be forced to rise, it will encounter lower pressure and expand. As it expands, its temperature will decrease. Conversely, if air should sink, it is compressed by higher pressure at lower levels and its temperature will increase. This change in temperature is due only to expansion or compression. No heat has been added to the air or subtracted from it. This type of process is called "Adiabatic Heating" or "Adiabatic Cooling."(Figure 3-9).

**Figure 3-9 Adiabatic heating and cooling**

11. If the air is unsaturated, it will cool or warm at a rate of about 3°C for each thousand feet of ascent or descent. This is known as the "Dry Adiabatic Lapse Rate" (DALR) (Figure 3-10).

12. If air should rise and cool until the temperature falls to the dew point, condensation will occur and cloud will form. During condensation, latent heat of vaporization is released to the rising air, reducing the rate of cooling. This new rate of cooling is called the "Moist" or "Saturation" Adiabatic Lapse Rate (SALR) (Figure 3-11).

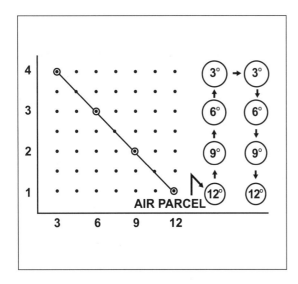

**Figure 3-10 Dry adiabatic lapse rate (DALR)**

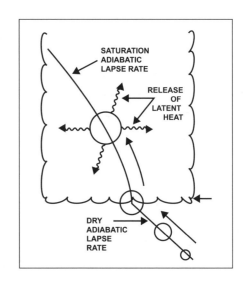

**Figure 3-11 Saturation adiabatic lapse rate (SALR)**

13. The amount of latent heat released during condensation depends on the amount of water vapour condensed. Since air can hold more water vapour at high temperatures, more latent heat is released when warm saturated air rises and is cooled than when cold saturated air is cooled. The moist adiabatic lapse rate varies from about 1.1°C/1,000 feet for warm air to 2.8°C /1,000 feet for cold air. An average of 1.5°C/1,000 feet is frequently used as an approximation.

14. If saturated air aloft should begin to sink, it will heat. As soon as it warms, it becomes unsaturated and will then heat at the dry adiabatic lapse rate. Rising air will cool at the dry adiabatic lapse rate until saturation is reached and will then cool at the moist adiabatic lapse rate. Subsiding air, even if it is initially saturated, will heat at the dry adiabatic lapse rate (Figure 3-12). When rain is falling through subsiding air, the air will heat at the saturated adiabatic lapse rate rather than the dry adiabatic lapse rate. In this case, heat is taken from the air to evaporate the falling rain (latent heat of vaporization) so as it subsides the air warms at the saturated adiabatic rate.

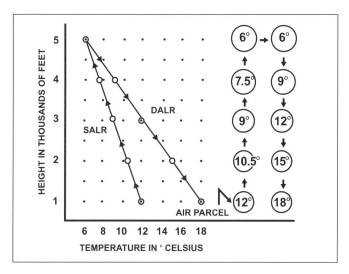

**Figure 3-12 Ascending and descending moist air**

# EXPANSION COOLING

## Orographic Lift and Upslope Lift

15. Air is forced to rise in several ways. "Orographic Lift" occurs when air flows against some topographical feature such as a mountain range. The layer of air that is lifted by this means extends above the obstruction, but the amount of lift decreases with altitude until a smooth flow occurs (Figure 3-13).

16. When there is a gradual increase in altitude as occurs from east to west across the Prairies, the term used to describe the lift is called "Upslope."

## Mechanical Turbulence

17. Wind blowing over rough terrain breaks up into whirls and eddies. In the resultant mixing, air from near the surface is lifted to higher levels and undergoes expansion cooling.

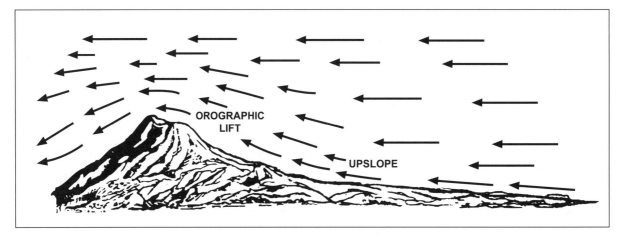

**Figure 3-13 Orographic lift and upslope lift**

**Figure 3-14 Mechanical mixing**

## Convection

18. "Convection Currents" are very localized ascending currents of buoyant air. As the current ascends, it expands and cools (Figure 3-15).

## Evaporation

19. Rain partially evaporates as it falls through the air. The heat required for the evaporation is taken from the air through which the rain is falling thus cooling the air (Figure 3-16).

## Advective Cooling

20. The lower levels of the atmosphere will be cooled if the air moves over a surface of the earth colder than itself. The cooling occurs due to conduction with mechanical turbulence mixing the cooling effect through a shallow layer of the air.

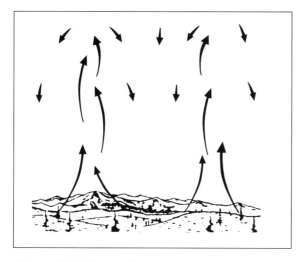

**Figure 3-15 Convection**

**Figure 3-16 Evaporation cooling**

## Large-Scale Ascent

21. The expansion cooling that has been described in the previous few paragraphs occurs over a relatively small area of the earth's surface, or within a fairly shallow layer of the atmosphere. There are situations in the atmosphere where the air over a large area, the size of Manitoba for example, slowly rises. Although the rate of ascent is very slow, in the order of a few feet per minute, it still causes expansion cooling. For instance, if the flow of air over an area converges, the air in the convergent zone is forced to ascend (Figure 3-17).

22. Fronts will be described in considerable detail later since they are major producers of cloud and precipitation. They also are areas of large-scale ascent where the air undergoes expansion cooling (Figure 3-18).

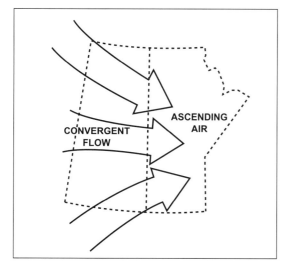

**Figure 3-17 Convergence**

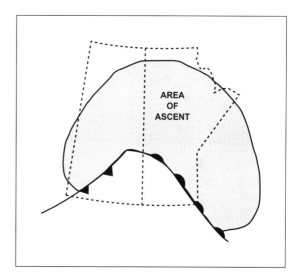

**Figure 3-18 Frontal ascent**

## Summary - Chapter 3

- A lapse rate is the rate of decrease of temperature with increased altitude.

- The environmental lapse rate is the change of temperature with altitude at any given time.

- An inversion is an increase of temperature with increased altitude.

- Radiation cooling from the earth causes a nocturnal or radiation inversion to form in the atmosphere at low levels at night.

- The intensity of the inversion is:
  - weak during windy conditions, under cloud cover, or in maritime areas;
  - strong under calm conditions and in valleys.

- Air undergoes adiabatic cooling when it rises and expands, and adiabatic heating when it subsides and compresses.

- Ascending or descending dry air changes temperature at the dry adiabatic lapse rate of 3°C/1,000 feet.

- Ascending saturated air changes temperature at the moist adiabatic lapse rate which averages 1.5°C/1,000 feet.

- Descending moist air changes temperature at the dry adiabatic lapse rate because it becomes unsaturated.

- Evaporation from rain falling through subsiding air can cause the air to remain saturated and warm at the saturated adiabatic lapse rate rather than at the dry adiabatic lapse rate.

- Expansion cooling occurs due to orographic or upslope lift, mechanical turbulence and convection.

- Cooling also occurs due to evaporation of rain.

- Large-scale ascent causing expansion cooling occurs with convergence and with fronts.

- Advection over a cold surface will cool a shallow layer.

# Chapter

# 4

There are times when you are flying that the air is completely smooth and not even a ripple is felt. There are other times when flying is so turbulent that it is difficult to control the aircraft. This difference is due to variations in the stability of the air.

# CHAPTER 4

## STABLE AND UNSTABLE AIR

### THE MEANING OF STABILITY IN RELATION TO AIR

1. A general idea of what the terms stability and instability mean in relation to air can be obtained from Figures 4-1 and 4-2. Figure 4-1 illustrates a ball in three different positions. In position 1, the ball is in a stable condition. If moved, it will return to its original placement. In position 2, the ball is in an unstable condition. It will not remain where it is. In the third position, the ball is in a neutral condition. If moved, it will remain in its new position, otherwise it will stay where it is.

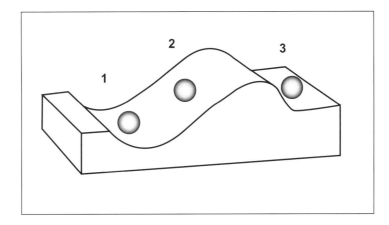

**Figure 4-1 Stability and instability**

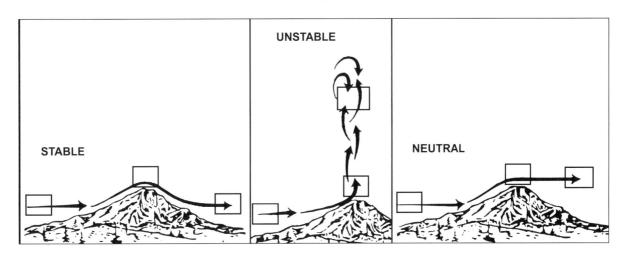

**Figure 4-2 Reaction of stable and unstable air to lift**

2.  Stability in the air relates to vertical movements of air "parcels." These parcels range in size from small turbulent eddies up to those measuring millions of cubic feet in volume. Figure 4-2 illustrates the behaviour of air with different stability characteristics as it encounters a mountain. In stable air, a parcel will rise up over the mountain then sink back down to its original level on the other side. In air with neutral stability a parcel will rise up the mountain then flow away at the mountain top altitude. In unstable air a parcel will continue rising vertically after striking the mountain. These airflows can be complicated by other factors, but this diagram does illustrate the basic differences in the various forms of air stability.

## Why Air is Stable or Unstable

3.  The action of the air parcels depends on their buoyancy within the atmosphere in which they are embedded. Since warm air is less dense than cold air, parcels that are warmer than the surrounding air will float up through it. A hot air balloon, for example, depends on this principle.

4.  The stability or instability of the air depends on the relationship between the temperature of the rising air parcel and the temperature of the surrounding air through which it is rising. The temperature of the surrounding air is indicated by the environmental lapse rate (ELR), an example of which is shown in Figure 4-3 where the temperature has been measured at various altitudes and plotted on a graph.

5.  You will recall from the previous chapter that a parcel of rising unsaturated air cools at the dry adiabatic lapse rate of 3°C/1,000 feet. If a parcel of air at a surface temperature of 4°C, as in Figure 4-3, is forced to rise to 1,000 feet, it will cool to 1°C. Since it is now colder and denser than the surrounding air, it will tend to sink back to the earth. This is stable air.

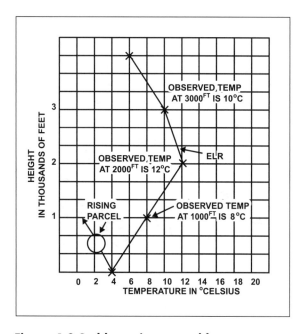

Figure 4-3 Stable environmental lapse rate

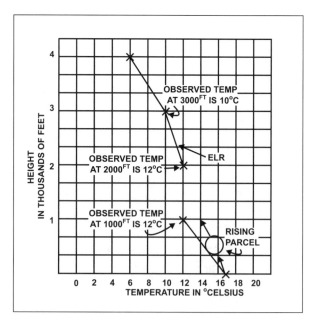

Figure 4-4 Unstable environmental lapse rate

6.  In Figure 4-4, a parcel with a surface temperature of 17°C will cool to 14°C if lifted to 1,000 feet. It is now warmer and less dense than the surrounding air, and will continue to rise until such time as it encounters air at its own temperature. This is unstable air.

7.  It can be seen from this that unsaturated air is unstable if the environmental lapse rate is greater than the dry adiabatic lapse rate as seen in Figure 4-4, and it is stable if the environmental lapse rate is less than the dry adiabatic lapse rate as seen in Figure 4-3. It has neutral stability if the environmental lapse rate is the same as the dry adiabatic lapse rate.

8.  If the air parcel rises until it becomes saturated, it will begin to cool at the saturated adiabatic lapse rate with further ascent. The air's stability will then depend on the relationship between the environmental lapse rate and the saturated adiabatic lapse rate rather than the dry adiabatic lapse rate. In this case the air will have neutral stability if the environmental lapse rate is the same as the saturated adiabatic lapse rate (Figure 4-5).

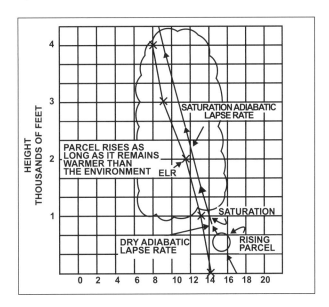

**Figure 4-5 Unstable saturated air**

## Descriptive Terms

9.  There are several terms used to describe the environmental lapse rate and the stability of the air:

  •  STEEP LAPSE RATE: the temperature decreases very rapidly with height. This implies unstable air.
  •  SHALLOW LAPSE RATE: the temperature decreases very little with height. This implies stable air.
  •  INVERSION: temperature increase with height; this indicates extremely stable air.
  •  ISOTHERMAL LAYER: the temperature does not change with height; this indicates very stable air.
  •  ABSOLUTE STABILITY: the environmental lapse rate is less than the saturation adiabatic lapse rate (SALR).
  •  ABSOLUTE INSTABILITY: the environmental lapse rate is greater than the dry adiabatic lapse rate (DALR).
  •  CONDITIONAL INSTABILITY: the environmental lapse rate is between the dry and the saturation adiabatic lapse rates. If the air is unsaturated, it is stable; if it is saturated, it is unstable.

- POTENTIAL INSTABILITY: Initially stable air becomes unstable as the whole air mass undergoes large-scale ascent until it becomes saturated; this would occur principally with frontal lift and with convergence.

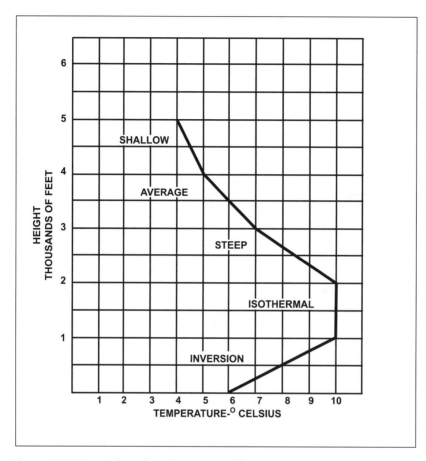

**Figure 4-6 Examples of environmental lapse rates**

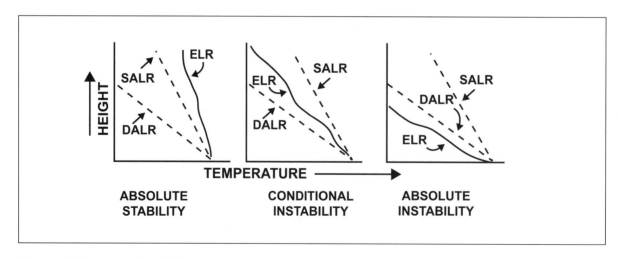

**Figure 4-7 Degrees of stability**

# Development of Stability and Instability

10. The stability or instability of the air depends upon its environmental lapse rate. If this lapse rate becomes steeper, instability can develop. If it becomes shallower, stability can develop.

11. The lapse rate of an atmospheric layer can be steepened by warming the lower portion of the layer or by cooling the upper portions. It can be made shallower by cooling the lower portion or by warming the upper portion (Figure 4-8).

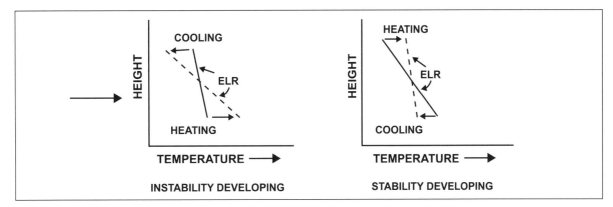

**Figure 4-8 Development of stability and instability in a layer of the atmosphere**

## Daytime Heating

12. Daytime surface heating is one of the principal methods of developing instability. The amount of heating depends on the type of surface, with water or wet soil heating only a small amount but dry soil heating intensely. The daytime heating can be so intense that a lapse rate greater than the dry adiabatic lapse rate develops in the lowest few hundred feet. This is called a "Super Adiabatic Layer." Night-time cooling of the surface develops stability in the lower layers. An inversion in the lower thousand feet or so develops but this again depends on the type of surface. Water areas cool little but dry land cools to a larger degree (Figure 4-9).

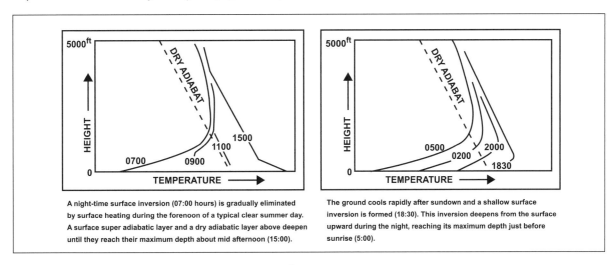

**Figure 4-9 Daytime heating and night-time cooling**

13. In the Arctic winter, air undergoes extreme cooling for an extended period of time. This cooling works up through the entire troposphere with the result that the air becomes basically stable throughout its entire depth. In the tropics, the air undergoes intense surface heating. This also works up to the tropopause and results in this air becoming basically unstable throughout its depth (Figure 4-10).

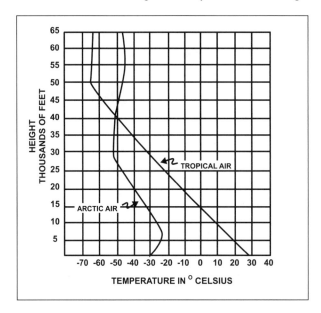

**Figure 4-10 Typical environmental lapse rates in arctic and tropical air**

14. The temperature structure in the stratosphere is generally isothermal or increases with height. For this reason, the stratosphere is always stable.

## Advection

15. The lower layers of the atmosphere can also be heated and made unstable if the air is carried by wind over a portion of the earth's surface that is warmer than itself. If the air should be carried over a surface colder than itself, it will be made stable (Figure 4-11).

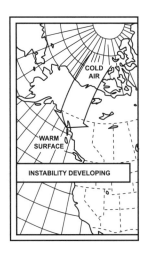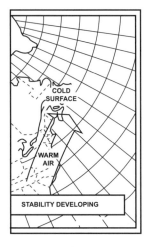

**Figure 4-11 Stability modification due to air movement**

## Cold Air Advection and Warm Air Advection

16. Changes in the temperature at the top of a layer in the atmosphere also occur and these alter the stability of the atmosphere below. In Figure 4-12 (a) cold air aloft is moving rapidly eastward and will develop instability in the layer of air at "X". Conversely, in Figure 4-12 (b), the warm air replacing the cold air will develop stability in the layer below it at "X". These processes are respectively "Cold Air Advection" and "Warm Air Advection."

## Subsidence Inversion

17. A very stable layer, or even an inversion, can form as a result of widespread sinking of air (subsidence) within a relatively thick layer aloft, while the air below this level remains unchanged. The sinking air is heated by compression (Figure 4-13).

## Convective Cells

18. The vertical currents that develop in unstable air are convective cells that consist either of a shaft of rapidly rising air surrounded by slowly settling air or of a large, buoyant bubble rising up through the atmosphere. These convective cells can occur at any height in the troposphere where the air is unstable, and in either clear air or embedded within widespread areas of cloud (Figures 4-14, 4-15).

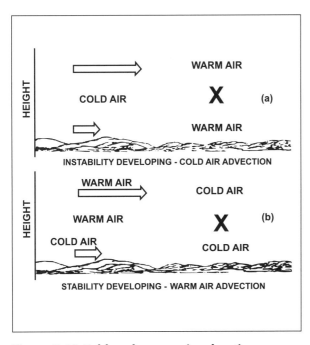

**Figure 4-12 Cold and warm air advection**

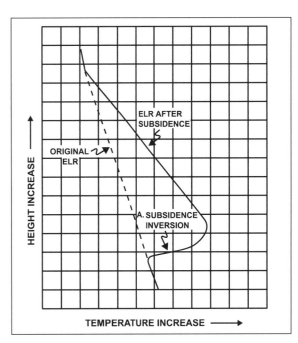

**Figure 4-13 Inversion aloft due to subsiding air**

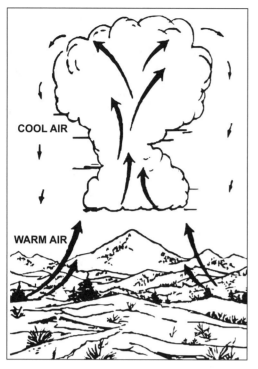

**Figure 4-14 Convective cell in clear air**

**Figure 4-15 Convective cell embedded in cloud**

## The Importance of Stability and Instability to Flying

19. Stability or instability is a basic characteristic of air that affects flying in many ways. The essential difference between them is that stable air hugs the earth's surface, or if it is a stable layer aloft, it acts as a lid to prevent vertical currents from rising through it, whereas unstable air bubbles or boils upwards, sometimes with surprising force. As you progress further through this book, you will learn in considerable detail the many important results caused by this difference. Some of these are summarized below.

### Visibility

20. Smoke and industrial pollutants from fires and cities are a major cause of poor visibility areas. Under stable conditions, they are trapped near the earth's surface and can cause severe restrictions to visibility. Under unstable conditions the smoke and pollution is mixed through the unstable layer so the visibility reduction is less.

### Cloud and Precipitation

21. In stable air, cloud forms in layers and in unstable air, it forms in towers. Precipitation is of a continuous type in stable air and of a showery type in unstable air. Fog, too, is characteristic of stable air.

## Wind

22. The wind at different altitudes can vary a great deal in stable air. If there is an inversion it can be markedly different above than below the inversion. With unstable air, the change of wind with altitude is not so marked. The wind tends to be gusty in unstable air and to blow at a steady speed in stable air. Topography affects the wind to a more marked extent in stable air so that hills and valleys and mountains influence the wind more in stable air than in unstable air.

## Turbulence

23. Flight in stable air is generally smooth although turbulent eddies can form in the zone between differing streams of air such as at the top of an inversion. Flight in unstable air is turbulent and, at times, extremely so.

**STABLE AIR**

Clouds in Layers, No Vertical Motion

Smoke Columns Flatten Out After Limited Rise

Poor Visibility in Lower Levels due to Accumulation of Haze and Smoke

Fog Layers

Steady Winds

**UNSTABLE AIR**

Clouds Grow Vertically and Smoke Rises to Great Heights

Towering Type Clouds

Upward and Downward Currents

Gusty Winds

Good Visibility

**Figure 4-16 Characteristics of stable and unstable air**

## Anomalous Propagation

24. The distance that radio transmissions of wavelengths less than about 10 metres can be received is limited to the distance of the radio horizon. These waves undergo a small amount of downward refraction by the atmosphere so the radio horizon is about one third greater than the visual horizon (Figure 4-17). This type of radio transmission includes VHF, UHF and radar.

25. Under certain atmospheric conditions, the path followed by the radio wave is significantly different from that normally expected and "Anomalous Propagation" occurs. This is particularly significant for radar transmission, and, when it occurs, targets can be detected at phenomenally long ranges. Anomalous propagation occurs with both land-based and airborne radars.

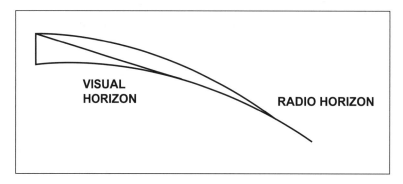

**Figure 4-17 Visual and radio horizons**

26. Under conditions of a strong inversion and a marked humidity decrease with height these wavelengths are refracted to a greater extent than normal. They are bent down until they strike the earth and then are reflected back up again. This process is repeated many times as shown in Figure 4-18. The layer where this occurs is called a radio duct. The signal strength is maintained and if the duct covers a wide area, targets many times the normal range of the radar can be detected.

27. The depth of the duct required for anomalous propagation varies from about 50 feet to around 1,000 feet and the direction of emission of the radio waves must be within about one half degree of the horizontal. Ducts can occur aloft as well as on the surface but are seldom above 15,000 to 20,000 feet.

28. The meteorological conditions favouring the formation of ducts include subsidence inversions over oceans, advection of warm, dry air over cooler water and nocturnal inversions. These all tend to combine a temperature increase and a moisture decrease with height. Unstable air is unfavourable for duct formation.

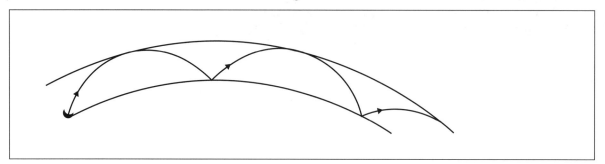

**Figure 4-18 A radio duct**

## Summary - Chapter 4

- Convective currents composed of air rising rapidly in shafts or as bubbles develop in unstable air.

- This type of motion is dampened out in stable air.

- There are various degrees of stability or instability that depend on how much the temperature of the atmosphere decreases with altitude. The more rapidly it decreases, the more unstable the air.

- Terms used to describe the environmental lapse rate are: steep, shallow, inversion, isothermal, average.

- Terms used to describe stability are: absolute stability, absolute instability, conditional instability, potential instability.

- Instability is developed in a layer by that layer being heated in the lower levels or cooled in the upper levels.

- Stability is developed in a layer by that layer being cooled in the lower levels or warmed in the upper levels.

- Flight characteristics of stable air:
  - Poor low-level visibility.
  - Layer cloud, continuous precipitation.
  - Steady winds that can change markedly with height.
  - Smooth flying.

- Flight characteristics of unstable air:
  - Good visibility.
  - Heap-type cloud, showery precipitation.
  - Gusty winds.
  - Turbulent flying.

- Anomalous propagation of radar may occur with a temperature inversion, and a moisture decrease with height.

# Chapter 5

Pressure patterns are a clue to wind, weather causes and the movement of weather systems. During flight you will routinely have to make adjustments for these.

# CHAPTER 5

# ATMOSPHERIC PRESSURE AND AIR CIRCULATION

1.   Atmospheric pressure is particularly important to aviation because it is used in aircraft altimetry to indicate the altitude of an aircraft. There are several indirect effects of pressure that are also important. The pressure distribution in the atmosphere controls the winds and, to a considerable extent, the occurrence of clouds and precipitation or clear skies. Air density is of considerable importance to aircraft performance and it is also controlled, in part, by atmospheric pressure. This chapter will describe various aspects of atmospheric pressure and the air movement resulting from it.

## Pressure

2.   "Atmospheric Pressure" at any level in the atmosphere is the force per unit area exerted by the weight of the air lying above that level.

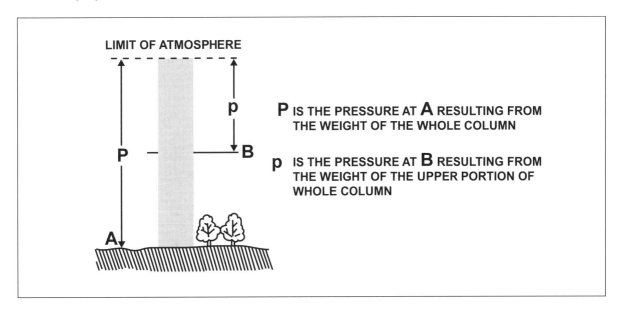

**Figure 5-1 Atmospheric pressure**

## Units and Methods of Measuring Pressure

3.   Pressure can be expressed in various units. Those that are used in aviation are inches of mercury, as used in altimetry, and hectopascals (hpa) as used in weather-map analysis. In public weather services, the kilopascal, which is 10 hectopascals (hPa) applies. Two common methods of measuring pressure are by digital barometer and by aneroid barometer.

4.   A "Digital Barometer" measures the atmospheric pressure electronically and displays the results digitally. It replaces the old mercury barometer.

5.   An "Aneroid Barometer" is a partially evacuated flexible metal cell that contracts with increasing pressure and expands with decreasing pressure. The change is registered on a scale by means of a needle and coupling mechanism. Either type of barometer can be calibrated in inches of mercury or in hectopascals.

## Station Pressure and Mean Sea Level Pressure

6.   When the pressure is measured at an airport, it is the weight of the air above the airport that is measured. This is called "Station Pressure." Pressure always decreases with height so the pressure at a high elevation will be less than that at a low elevation.

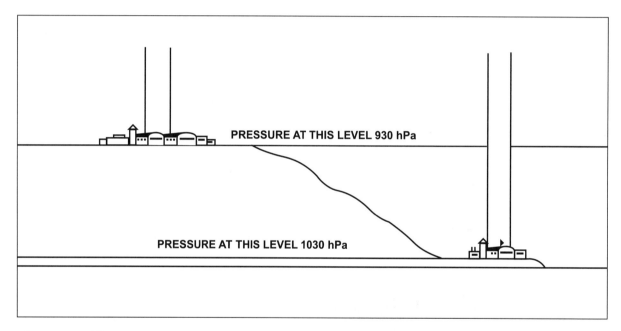

PRESSURE AT THIS LEVEL 930 hPa

PRESSURE AT THIS LEVEL 1030 hPa

**Figure 5-2 Difference in station pressure due to difference in station elevation**

7.   In order to analyze weather maps, the pressure at different observing stations must be compared. Since pressure varies with the station elevation, these pressures must be adjusted to some common level to make the comparison. The level used is "Mean Sea Level" (MSL).

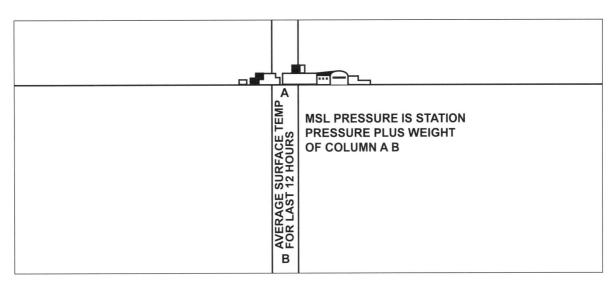

**Figure 5-3 Mean sea level pressure**

8.  The mean sea level pressure at station A is the station pressure plus the weight of the fictitious column of air between the station and mean sea level. The weight of this column is dependent on its temperature, which is assumed to be the average temperature at the station over the last 12 hours.

## Isobars

9.  To provide a visual portrayal of the pressure patterns across the country, the mean sea level pressure from the observing stations is plotted on a chart called a "Surface Weather Map." Lines called "Isobars" are drawn on this chart joining places having the same MSL pressure. They are drawn at four hectopascal intervals up and down from 1,000 hectopascals. These lines never cross and form pressure patterns that are related in a complex way to the weather that is occurring.

## Pressure Systems

10. Figure 5-4 is a sample of isobaric patterns on a surface chart. In this example, isobars have been drawn at four-hectopascal intervals from 1,000 hectopascals up to 1,024 hectopascals. They have divided the country into areas having different pressure systems.

    a.  LOW PRESSURE AREAS - These are also called "Depressions" or "Cyclones." They are areas of MSL pressure surrounded on all sides by higher pressure and are marked by an "L" on surface charts. The term "Cyclonic Curvature" is sometimes used and it is the curvature of isobars to the left if you were to stand with lower pressure to your left. Area A in Figure 5-4 is an area of cyclonic curvature.

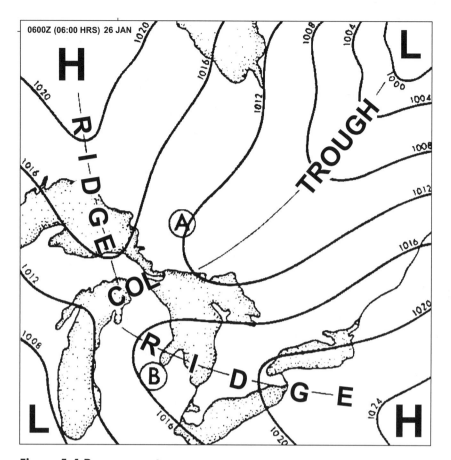

**Figure 5-4 Pressure systems**

b.  HIGH PRESSURE AREAS - These can also be called "Anticyclones.. They are areas of pressure surrounded on all sides by lower pressure and are marked by an "H" on surface charts. "Anticyclonic Curvature" is the curvature of isobars to the right if you were to stand with lower pressure to your left. Area B in Figure 5-4 is an area of anticyclonic curvature.

c.  TROUGHS - These are elongated areas of low pressure with the lowest pressure along the line of maximum cyclonic curvature.

d.  RIDGES - These are elongated areas of high pressure with the highest pressure along the line of maximum anticyclonic curvature.

e.  COLS - These are neutral areas between two highs and two lows as shown in Figure 5-4.

11.  You should note that all these pressure systems are relative to the pressure surrounding them. The area over Nova Scotia with a 1,008 hectopascal isobar around it in Figure 5-5(a) is a high pressure area because the pressure surrounding it is less than 1,008 hectopascals. In Figure 5-5(b) this same area, still with a 1,008 hectopascal isobar around it, but at a different time, is now a low pressure area because the pressure surrounding it is higher than 1,008 hectopascals.

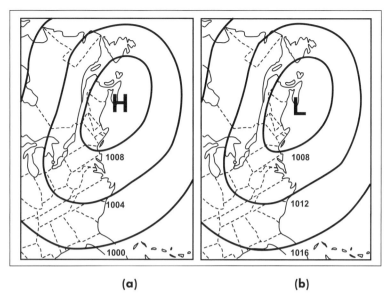

**Figure 5-5 Pressure centres in relation to the surrounding pressure**

## Pressure Tendancy

12. The pressure systems outlined on surface charts are not constant, but rather increase and decrease in intensity and drift across the country. This means that the pressure at any particular point is seldom steady and is usually increasing or decreasing at amounts dependent on how fast the systems are moving or changing intensity. The rate of rise or fall of pressure at a particular location is called the "Pressure Tendency."

# Air Circulation

## Pressure Gradient Force

13. If there is a pressure difference across the country, air will begin to move from the region having high pressure directly towards the area with low pressure. The force causing this movement is called the "Pressure Gradient Force" (PGF) and its strength is dependent on the pressure difference over the area. On a surface chart, the extent of the pressure difference can be seen by the spacing of the isobars. If the isobars are spaced closely together there is said to be a "Steep" or "Strong" pressure gradient. If they are far apart the pressure gradient is described as "Weak" or "Flat."

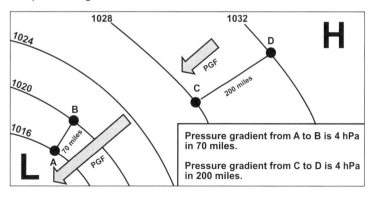

**Figure 5-6 Pressure gradient**

## Coriolis Force

14. As soon as the air begins to move, it is influenced by another force called the "Coriolis Force" (CF). This is a complicated force that is a result of the earth's rotation. It causes air in motion to deflect to the right in the northern hemisphere and to the left in the southern hemisphere. The strength of the Coriolis force increases with increased air speed and also varies from zero at the equator to a maximum strength at the poles.

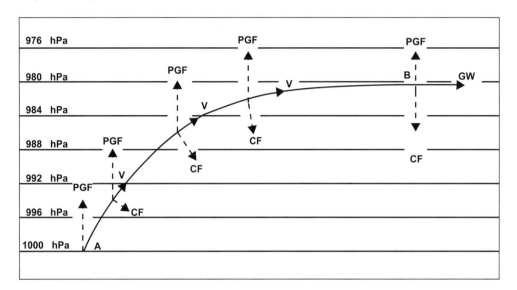

**Figure 5-7 The geostrophic wind**

## Geostrophic Wind

15. In Figure 5-7, a parcel of air at A under the influence of the pressure gradient force begins to move towards a lower pressure. As it moves, the Coriolis force deflects it to the right. Under the continued influence of the pressure gradient force, the speed of the parcel increases and this causes an increase of the Coriolis force. Eventually, as at B, the Coriolis force has increased to a value that just balances the pressure gradient force. A state of equilibrium has been reached with both forces equal and opposite and the air motion steady and parallel to the isobars. The resulting wind is called the "Geostrophic Wind." The stronger the pressure gradient, the greater the resulting geostrophic wind speed.

## Buys-Ballot's Laws

16. The effect of the Coriolis force is expressed in Buys-Ballot's Law:

*"If you stand with your back to the wind in the northern hemisphere, low pressure will be on your left."*

This means that air flows clockwise around a high and counterclockwise around a low, in the northern hemisphere. The reverse is true in the southern hemisphere.

# Non-Geostrophic Winds

## Latitude Effect

17. There are several forces that act on air in motion that influence the extent that the motion is geostrophic. The first of these originates from the fact that the Coriolis force varies from zero at the equator to a maximum at the poles. This means that from about 15° north to 15° south, air does not flow parallel to the isobars, but tends to flow more directly from high to low pressure.

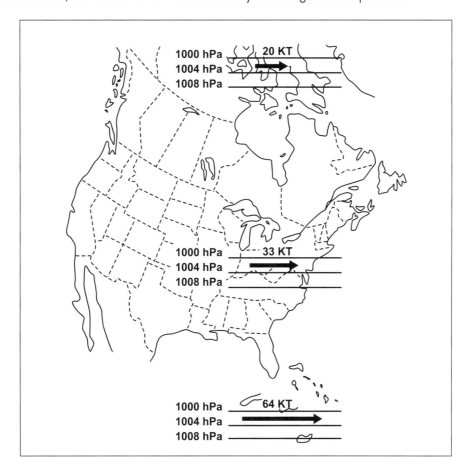

**Figure 5-8 Latitude effect on geostrophic wind**

18. The wind is geostrophic when the Coriolis force balances the pressure gradient force. The Coriolis force is greatest near the poles and increases as wind speed increases. This means that the Coriolis force will require less wind to balance the pressure gradient force in polar regions. For example, the pressure gradient or isobar spacing which produces a wind of 20 knots at 60°N will produce a wind of 33 knots at 40°N and 64 knots at 20°N. This variation in wind speed with isobar spacing should be kept in mind when you are looking at weather charts that cover an extensive area of the earth.

## Curvature Effect

19. A geostrophic flow occurs only when the isobars are straight. When the isobars are curved, the air moves in an arc or a circle and "Centrifugal Force" (cf) comes into play. Figure 5-9(a) illustrates a flow with straight isobars. The pressure gradient force is from high to low, and the air flow is to the south such that high pressure is to the west. Figure 5-9(b) illustrates a cyclonically curved flow. The centrifugal force is acting in opposition to the pressure gradient force, so the wind speed is less. Figure 5-9(c) illustrates an anticyclonic circulation. The centrifugal force is acting in the same direction as the pressure gradient force so the wind speed is greater. For the same pressure gradient, the wind speed around a low is less than around a high. The pressure gradient around lows, however, is generally very much stronger than around highs so that stronger winds normally occur with lows.

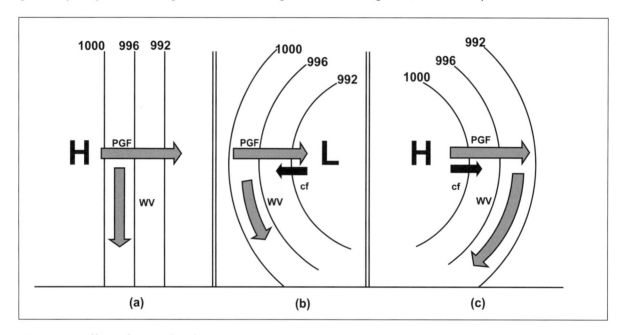

**Figure 5-9 Effect of curved isobars**

## Friction Effect

20. Topographical features on the earth's surface cause friction that tends to retard air movement and reduce the wind speed in the lower levels. Since the Coriolis force varies with the wind speed, a reduction of the wind speed will reduce the Coriolis force. With the pressure gradient force constant, a reduction of the Coriolis force will cause the wind to angle across the isobars into low pressure. This angle varies on the amount of friction imposed by the earth's surface and would be about 10° over oceans and 40° over very rough terrain (Figure 5-10). Frictional effects are greatest near the ground, but the effects are also carried aloft by mixing of the air. Air at around 2,000 to 3,000 feet above ground can be considered to be free of friction effect and to flow parallel to the isobars with a speed proportional to the pressure gradient (Figures 5-11, 5-12).

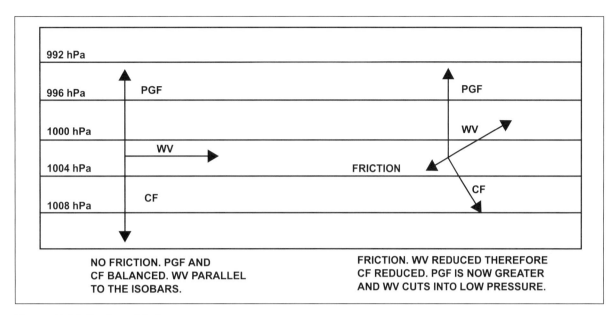

NO FRICTION. PGF AND
CF BALANCED. WV PARALLEL
TO THE ISOBARS.

FRICTION. WV REDUCED THEREFORE
CF REDUCED. PGF IS NOW GREATER
AND WV CUTS INTO LOW PRESSURE.

**Figure 5-10 Surface friction**

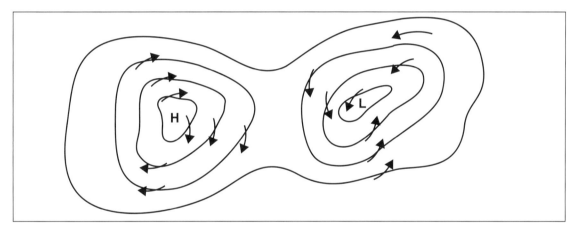

**Figure 5-11 Wind direction around isobars (surface)**

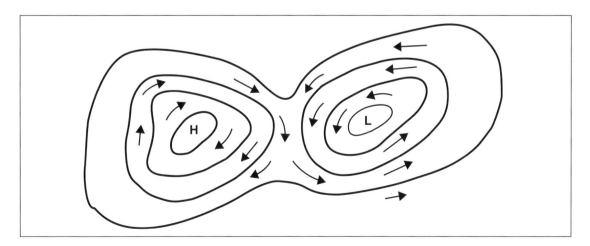

**Figure 5-12 Wind direction around isobars (2,000-3,000 feet)**

## Changing Pressure

21. Although intensifying or weakening pressure systems cause only a small departure from geostrophic winds, they are of great importance in the development of clouds and precipitation. If a low is deepening, the central pressure is falling and the pressure gradient force is increasing. A balanced condition with the Coriolis force does not exist, and this, in addition to friction, causes a slight cross-isobar flow into the low pressure centre with the air spiralling slowly upwards. This is the primary cause of the extensive areas of cloud and precipitation that occur with deepening low pressure areas. Similarly, in an intensifying high, the pressure is rising and the pressure gradient force is increasing. This again causes a slight cross-isobar flow out of the high pressure area with the air this time slowly descending over the high.

## Provision of Wind Information for Aviation

22. For aviation, the Weather Service provides the wind speed in knots and the direction in degrees true for both surface conditions and for winds aloft. Air traffic services will normally provide the local aerodrome surface wind in degrees magnetic. The direction is the direction from which the wind is blowing. For example a 270° wind is a wind blowing from the west.

## Terms Describing Wind

23. The wind may have a smooth, steady flow, or it may contain "Gusts" or "Squalls." In a gusty flow, there are rapid peaks and lulls in the wind speed; in a squall there is a sudden increase lasting for a minute or more, then a decrease. Gusts and squalls imply turbulent flight and you will be advised if the wind has these characteristics.

24. The terms "Veer" and "Back" are sometimes used to describe a change in wind direction. A veer is a clockwise change in direction and a back is a counterclockwise change.

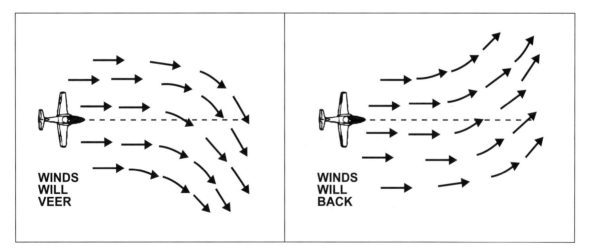

**Figure 5-13 Aircraft encountering veering and backing winds**

## Summary - Chapter 5

- Pressure at a given level is the force caused by the weight of the air above the level. It is measured by aneroid or digital barometers and the units used for aviation are hectopascals and inches of mercury.

- Station pressure is measured at the station elevation.

- Mean Sea Level Pressure is the station pressure plus the weight of the fictitious column from the station elevation to sea level.

- Isobars are lines drawn on surface charts joining places of equal pressure. They form patterns called lows, troughs, highs, ridges and cols.

- The pressure tendency is the rate of change of pressure at a location caused by the movement and change in intensity of the pressure systems.

- In the northern hemisphere, air flows counterclockwise around a low and clockwise around a high, parallel to the isobars with a speed proportional to the pressure gradient. This is valid only north or south of 15° latitude.

- Surface friction decreases the wind speed in the lower levels and causes the wind to cut in towards low pressure and out of high pressure.

- For a given pressure gradient, wind speed is less in polar regions than nearer the equator, and less around low pressure areas than high pressure areas.

- In a deepening low, the rate at which air spirals inwards and upwards increases, whereas in an intensifying high, the rate at which it spirals downwards and outwards increases.

- Wind speed is given in knots and direction in degrees true. If gusts or squalls are occurring, this information is provided.

- A veer is a clockwise change in wind direction; a back, a counterclockwise change in wind direction.

**5** CHAPTER

# Chapter

# 6

There are two basic types of weather – air mass and frontal. You will find that both can be severe.

# CHAPTER 6

# AIR MASSES AND FRONTS

## Air Masses

1. You have learned that there are various features of the atmosphere that have a control on the type and extent of the weather that occurs. These include the moisture content of the air, its temperature, stability and tropopause height. If a large body of air should stagnate over a portion of the earth's surface having uniform moisture and temperature, the air will acquire specific characteristics of these control features.

2. A body of air, usually 1,000 or more miles across, which has acquired uniform characteristics, is called an "Air Mass." Within horizontal layers, the temperature and humidity properties of an air mass are fairly uniform. Air from the surface mixes throughout the troposphere but does not penetrate the stratosphere. For this reason air masses are topped by the tropopause.

3. Although the moisture and temperature characteristics are uniform, the actual weather within an air mass may vary due to different processes acting on it in different areas. An air mass that is basically clear, for instance, may produce cloud in an area where it is undergoing orographic lift. Different air masses will develop different types of weather. In the instance of orographic lift just mentioned, layer cloud may develop in one air mass and thunderstorms in another. The type of weather depends on the characteristics of the air mass. (Figure 6-1)

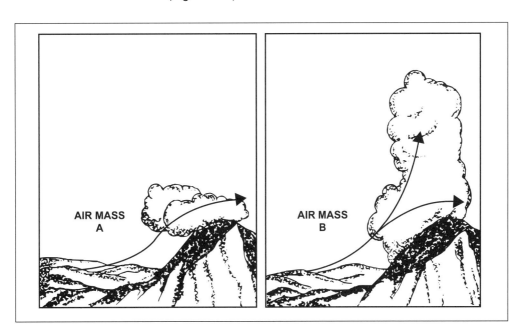

**Figure 6-1 Air mass weather**

4.  Slow moving high pressure areas provide prime conditions for the formation of air masses although they can also form if they move over a uniform surface for a long period of time. The regions where they acquire their characteristics are called "Source Regions." Ocean areas, snow or ice covered areas, desert areas and tropical areas are common source regions.

## Classification of Air Masses

5.  Because air masses tend to have their own characteristic weather, it is useful to be able to refer to them by a name. The names selected are based on their moisture content and temperature. Dry air is called Continental and moist air is called Maritime. The warmest air is called Tropical and the coldest air, Arctic. There is an air mass with intermediate temperatures called Polar.

6.  The names of the air masses and the abbreviations used to identify them on weather maps or in aviation forecasts are as follows:

| | |
|---|---|
| Continental Arctic | – cA |
| Maritime Arctic | – mA |
| Continental Polar | – cP |
| Maritime Polar | – mP |
| Continental Tropical | – cT |
| Maritime Tropical | – mT |

## Description of Air Masses

7.  The temperature, moisture content, stability and tropopause heights of the air masses are:

Continental Arctic — dry, very cold, very stable throughout, very low tropopause.
Continental Polar — dry, cold, fairly stable throughout, low tropopause.

### Note

The terrain features of North America are such that Continental Polar air is not found in Canada but it is found elsewhere in the world including the United States.

Maritime Arctic — moist, cold, unstable in the lower levels, low tropopause.
Maritime Polar — moist, cool, unstable, medium tropopause.
Maritime Tropical — moist, hot, very unstable, high tropopause.
Continental Tropical — dry, very hot, very unstable, very high tropopause.

### Note

The source area for Continental Tropical air is very small in North America, so this air mass is only occasionally found in Canada.

8.  In Figure 6-2 note that the warmer an air mass is, the higher and colder its tropopause. Note also that the temperature difference between air masses is greatest at low levels in the troposphere and least at high levels.

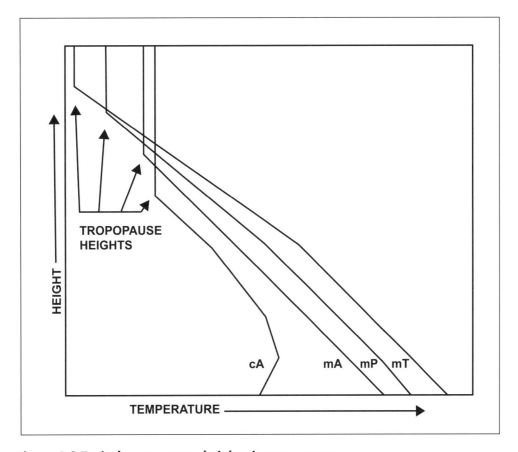

**Figure 6-2 Typical temperature–height air mass curves**

## Modification of Air Masses

9.  Air masses frequently move out of their source regions, and their characteristic features may change. The modification can be extensive enough so that a new air mass will form. The degree of modification depends on:

    a.  the speed with which they move;

    b.  the moisture or dryness of the region over which they travel; and

    c.  the temperature difference between the new surface and the air masses.

## Warming from Below

10. Warming from below develops instability and convection in the lower levels of the atmosphere. If this occurs, the earth's surface characteristics of dryness, moisture and temperature are carried aloft and modify the air mass to a considerable height. How high depends on how long the warming continues and how intense it is. In some cases, the entire air mass will be modified up to the tropopause.

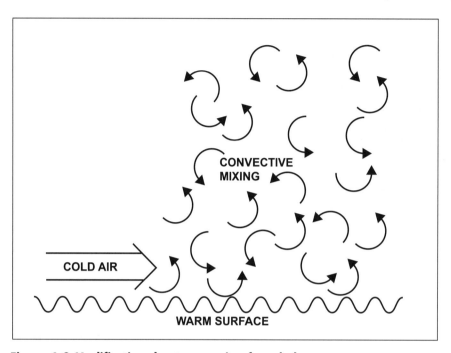

**Figure 6-3 Modification due to warming from below**

## Cooling from Below

11. Cooling from below causes increased stability in the lower levels. This blocks vertical motion so that any modifications of moisture and temperature occurring in an air mass that is moving over colder terrain will be restricted to the lower few thousand feet.

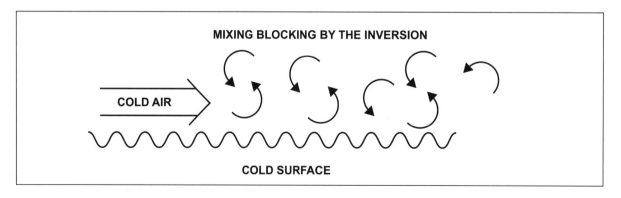

**Figure 6-4 Modification due to cooling from below**

## Sources and Modification of Air Masses

12    Figure 6-5 illustrates the source regions for the air masses over North America and their modification when they leave their source regions.

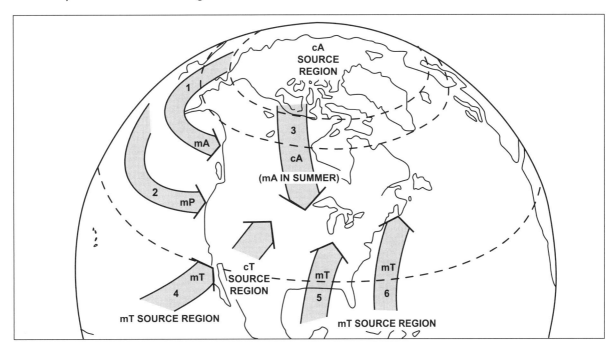

**Figure 6-5 Air mass source regions**

13. The Arctic area that is covered with snow and ice is the source for Continental Arctic air. In winter this is a very large area, but in summer it is much smaller and is restricted to the immediate polar area. This air then frequently breaks out of the Arctic and follows path 1 over the North Pacific. It is intensely heated from below so that vigorous convection develops and moisture from the ocean surface evaporates rapidly into the air. It is in convective turmoil in the lower levels and is very turbulent with frequent rain showers or snow showers. What was Continental Arctic air rapidly changes to Maritime Arctic air.

14. If this air should travel further south along path 2, where the water surface is warmer, the air finally reaches the temperature of the surface and the convective activity becomes more subdued. The temperature and moisture modification will have been carried throughout the air mass up to the tropopause. It is now Maritime Polar air.

15. If the Continental Arctic air moves out of its source region down the centre of the continent along path 3, its modification will depend on whether it is winter or summer. In winter, there is little modification as long as the air remains over a snow surface. South of the snow line, it will be heated from below and modify to Continental Polar air. These are the coldest wintertime air masses and they generally give clear weather except for localized areas near cities or open lakeswhere moisture is added to the air.

16. Northern Canada is covered with thousands of lakes so that in summer, when the lakes are not frozen, an outbreak of Arctic air following path 3 picks up moisture and is modified to Maritime Arctic air. It will be characterized by convective activity and rain showers.

17. Maritime Tropical air originates over the tropical waters of the Caribbean and the Pacific and typically follows paths 4, 5, or 6. The air is normally very warm, moist and unstable throughout its depth. In winter the air is cooled from below as it travels northward so that an inversion develops in the lower levels; however, there is little change in the air above the inversion. In the winter, this air mass seldom reaches as far north as Canada on or near the earth's surface, although it may be found at higher levels in the atmosphere.

18. In the summer, radiation cooling over land at night will cool the lower levels and cause an inversion, but daytime heating is sometimes sufficiently strong to burn this off so that the air mass is again very unstable and violent convection can occur. If the air should travel northward over cool water along path 6 in the summer, it will be cooled from below and an inversion will develop that has little diurnal variation.

19. The source region for Continental Tropical air is the desert area of the southern United States. This will never reach Canada in winter. In summer, it may occasionally move northward up the centre of the continent into southern Canada. Except for night-time inversions forming, little modification will occur.

## Fronts

20. Cold, dense air, does not mix readily with warm, less dense air. A cold air mass and a warm air mass lying adjacent to one another will mix only slightly at their border. The temperature of each air mass will be fairly uniform within itself, but there will be a large change of temperature within a relatively short distance of 50 to 100 miles in the zone between the air masses. The transition zones between air masses where the temperature changes rapidly are called "Fronts" or "Frontal Systems" (Figure 6-6)

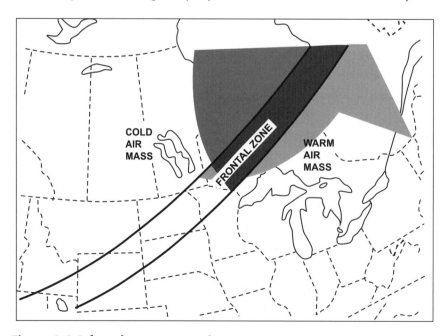

**Figure 6-6 A front between two air masses**

## Frontal Systems

21. The name of each frontal system is based on the name of the colder air mass involved. The common air masses in Canada from coldest to warmest are Continental Arctic, Maritime Arctic, Maritime Polar and Maritime Tropical. Figure 6-7 illustrates these four air masses with the frontal system between them. The frontal system between cA and mA air is called the "Arctic Front," between mA and mP the "Maritime Front" and between mP and mT the "Polar Front." On a weather map, they will be marked with an "A," an "M," or a "P." If there is no zone of abrupt temperature change, such as occurs when cA modifies to mA, there will be no front between the air masses.

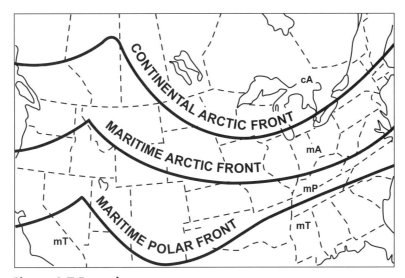

**Figure 6-7 Frontal system**

## Types of Fronts

22. Figure 6-8 illustrates a cold air mass travelling eastward across Canada. It is a big dome of cold air topped by the tropopause with its eastern edge advancing into Eastern Canada. The leading edge of an advancing cold air mass such as this is called a "Cold Front."

23. The cold air is retreating out of British Columbia. The retreating edge of a cold air mass is called a "Warm Front."

24. The cold air to the north or south of the air mass is neither advancing nor retreating. These are called "Quasi-Stationary Fronts."

25. A front is cold, warm or quasi-stationary depending on the motion of the cold air. On weather maps, cold fronts are marked in blue, or by solid triangles pointed in the direction that the front is moving. Warm fronts are marked in red, or by solid semi-circles pointed in the direction that the front is moving. Quasi-stationary fronts are marked in alternating red and blue lines or by alternating triangles and semi-circles with the triangles facing away from the cold air. Marking fronts on weather maps in this manner leaves the impression that there is an abrupt boundary between the air masses. You should always remember that, in fact, a "Front" is a zone several miles in extent where one air mass blends into another.

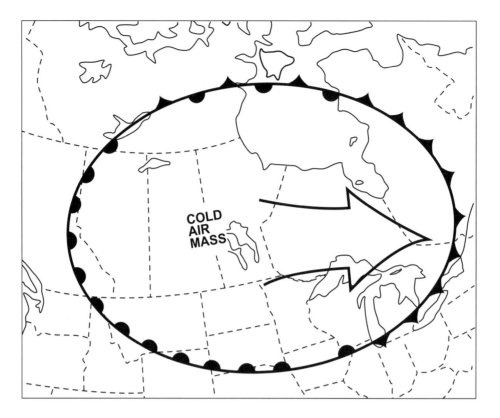

**Figure 6-8 Cold, warm and quasi-stationary fronts**

## Frontolysis and Frontogenesis

26. For a front to exist between two air masses, the temperature change from one to the other must occur within a relatively short distance. If the temperature contrast between the two air masses should decrease or if the zone where the temperature changes from one air mass to the other should become very broad, the front will disappear. This process is called "Frontolysis." It can occur either due to the modification of the air masses by the earth's surface, or due to a wind field that weakens the temperature gradient in the frontal zone.

27. "Frontogenesis" is the reverse process of frontolysis. If two air masses are lying adjacent to one another and the wind field is such that it tightens the temperature gradient in the zone between them, a front can form.

# Summary - Chapter 6

- Air masses are huge bodies of air having uniform characteristics of moisture and temperature in the horizontal. They lie within the troposphere.

- Air masses form in source regions when air stagnates over a portion of the earth's surface having uniform properties.

- The names of the air masses commonly found in Canada are:

  - Continental Arctic (cA)
  - Maritime Arctic (mA)
  - Maritime Polar (mP)
  - Maritime Tropical (mT)

- Air masses frequently migrate from their source regions. They may, or may not, become modified during migration depending on the type of surface over which they are moving.

- Air masses migrating over a warmer surface cause modification throughout the convective mixing layer, and modification to a new type of air mass is possible.

- Air masses migrating over a colder surface develop an inversion with modification only in the lower levels.

- The transition zones between air masses are called fronts.

- A cold front is the leading edge of an advancing cold air mass. On weather maps it is marked in blue or as a black line with solid triangles.

- A warm front is the trailing edge of a cold air mass. On weather maps, it is marked in red or as a black line with solid semi-circles.

- A quasi-stationary front is the edge of a cold air mass that is neither advancing nor retreating. It is marked on weather maps by alternating red and blue lines or as a black line with triangles and semi-circles.

- Frontolysis is the term used to describe the dissipation of a front.

- Frontogenesis is the term used to describe the formation of a front.

# Chapter

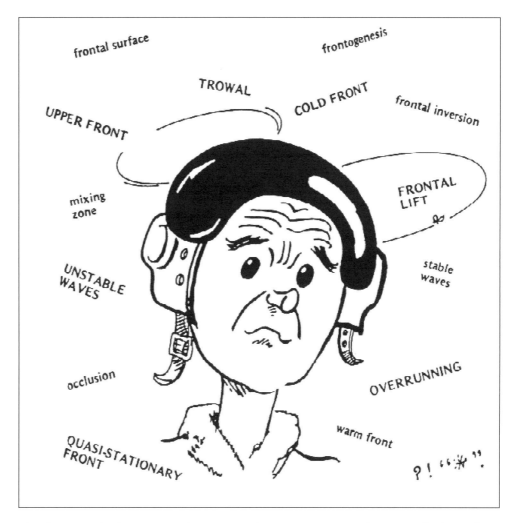

The front marked on a weather map indicates a structure that extends throughout the troposphere. This chapter will describe the important features of this structure.

# CHAPTER 7

# THE STRUCTURE OF FRONTS

## Mixing Zone

1. You will recall that a front is the transition zone between two air masses. In this zone, the air temperature changes from that prevalent in one air mass to that of the other in a belt roughly 50 to 100 miles across. The front is marked on a weather map by a line that is drawn along the warm-air side of the transition or mixing zone.

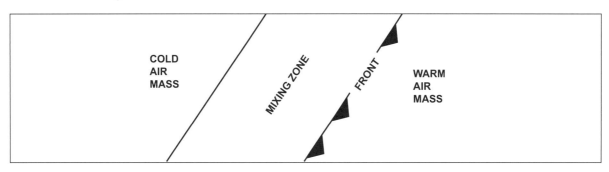

**Figure 7-1 Frontal position on the warm side of the mixing zone**

## Frontal Surface

2. A cold air mass lies in a big inverted saucer shape over the earth. The adjacent warm air mass lies all around and over the cold air. In Figure 7-2 a large pool of cold air has formed over Canada and the Arctic and it is surrounded and overrun by the warmer air.

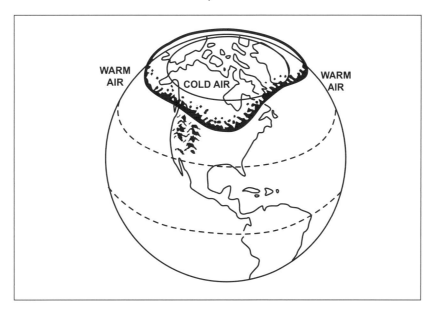

**Figure 7-2 A cold air mass surrounded by warmer air**

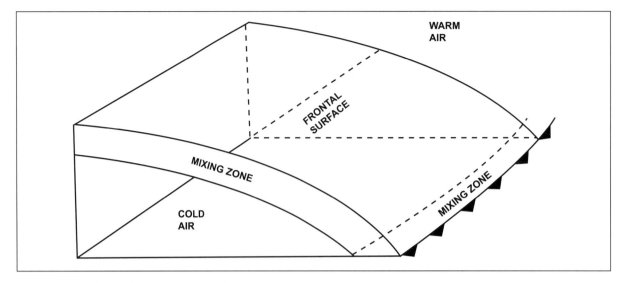

**Figure 7-3 A frontal surface**

## Frontal Surface And Frontal Inversion

3.   Figure 7-3 is a section of the cold air mass near the edge where is touches the earth's surface. The warm air side of the mixing zone on the earth's surface is marked as a cold front, and aloft, is called a "Frontal Surface." If you were to ascend vertically from the cold air into the warm air, there would be a temperature increase as you climbed through the mixing zone as shown in Figure 7-4. This temperature increase is called a "Frontal Inversion."

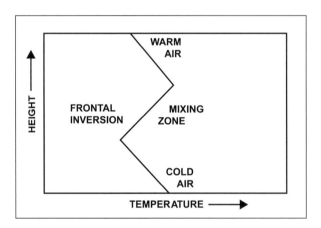

**Figure 7-4 A frontal inversion**

## Front Cross-Section

4.   Figure 7-5 presents the three-dimensional features of fronts by means of cross-sections. Figure 7-5(a) shows a cold air mass lying over North America outlined by a quasi-stationary front. Two cross sections through the atmosphere are provided in Figure 7-5(b) and (c). Along A-B-C-D the cold air is very shallow and is completely overrun by the warmer air. The warm air tropopause is shown with the stratosphere above it. Along E-F-G-H the cold air is very much deeper and has built up to reach the tropopause. You will recall that the tropopause over cold air is at a lower level than over warmer air so in this situation we have a warm air tropopause at a high level dropping down to a cold air tropopause at a lower level over the cold air.

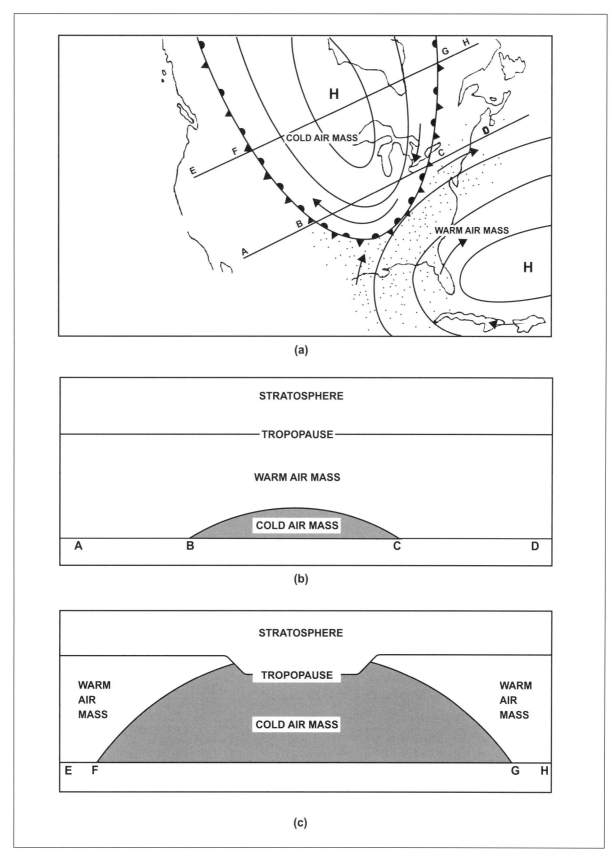

Figure 7-5 Cross-sections through frontal surfaces

5.  There are occasions when there is more than one cold air mass and one warm air mass. In these cases, the coldest air mass will lie as the lowest inverted saucer shaped shell, with the next coldest forming a shell over this and so forth until the warmest air mass is reached. Each air mass will normally be topped by the tropopause. Figure 7-6 illustrates a situation with the four air masses commonly found over North America and how a cross-section through these would appear.

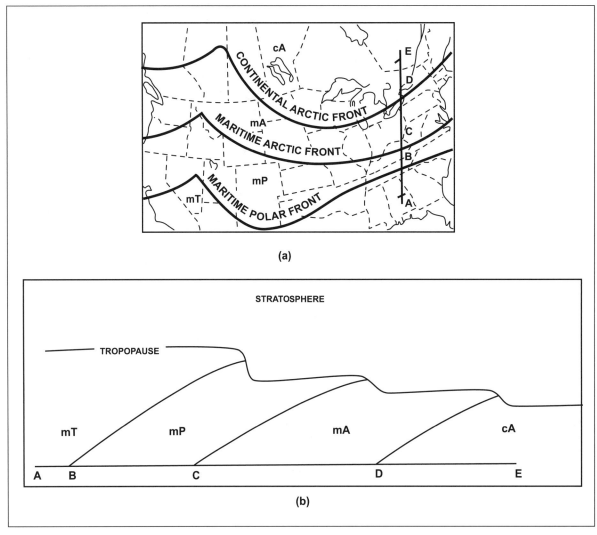

**Figure 7-6 Cross-section through four air masses**

## The Movements of Fronts

6.  Fronts were described earlier as cold, warm or quasi-stationary depending on the motion of the cold air. A cold front was described as the leading edge of a cold air mass and a warm front as the trailing edge of a cold air mass. In a quasi-stationary front the cold air is neither advancing nor retreating. The motion of the cold air is indicated on a surface map by the isobars in the cold air. Figure 7-7 illustrates the isobaric patterns for cold, warm and quasi-stationary fronts. Note that the front always lies in a trough of low pressure. In Figure 7-7(a) the wind speed in the cold air behind the cold front is 30 knots. The component of this wind perpendicular to the cold front is only 15 knots and it is this component perpendicular to the front that moves the front. In Figure 7-7(b) the wind speed in the cold air ahead of the warm front is 25 knots. The component of this wind perpendicular to the front is 10 knots and this is the speed of the front. In Figure 7-7(c) the cold air speed is 25 knots but its direction is parallel to the front so the front is quasi-stationary.

**Note**

The movement of the front is dependent on the motion of the cold air perpendicular to it. The motion of the warm air does not affect the movement of the front.

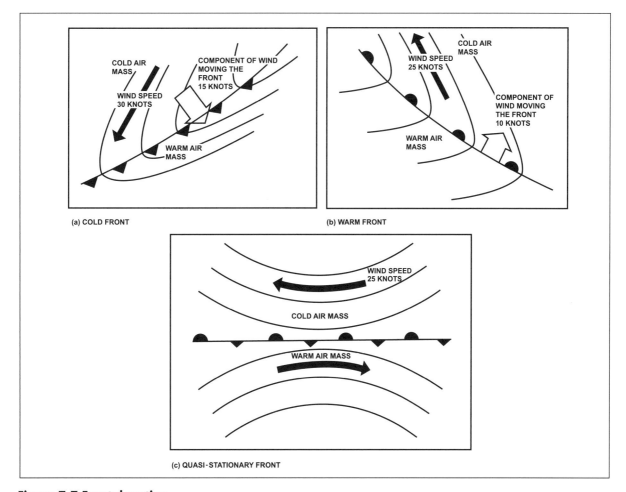

**Figure 7-7 Frontal motion**

## The Slope of Frontal Surfaces

7.    The slope that a frontal surface makes with the earth's surface is extremely shallow and is in the order of one degree. Surface friction modifies this slope. Figure 7-8(a) illustrates a stationary front with a slope of one degree. Note that the vertical dimension is exaggerated. This is done in all illustrations of fronts for increased clarity. Figure 7-8(b) is a warm front moving at 15 knots. Surface friction has caused a drag in the lower levels so that the slope has been reduced to one half degree. Figure 7-8(c) is a cold front moving at 20 knots. In this case, the surface drag has increased the angle to one and a half degrees. If the speed of a cold front is relatively high and the surface friction large, it can develop a protruding nose in the lower few thousand feet as shown in Figure 7-8(d). This is a very unstable air mass arrangement because cold, dense air is lying over warm, lighter air and violent weather called a "Line Squall" can develop.

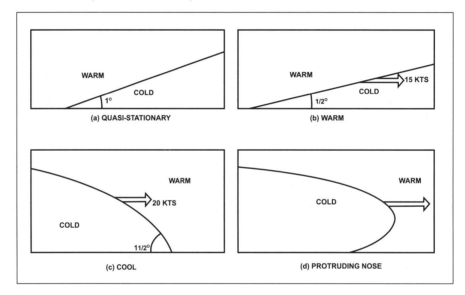

**Figure 7-8 Frontal surface slopes**

8.    The angles for the frontal slopes that have been given are examples only and will vary in different situations. The slope can also be given as a ratio. In Figure 7-9, the frontal slope has a ratio of 1 in 150. An aircraft flying at around 6,000 feet would encounter the frontal surface 150 nautical miles past the surface position of the front.

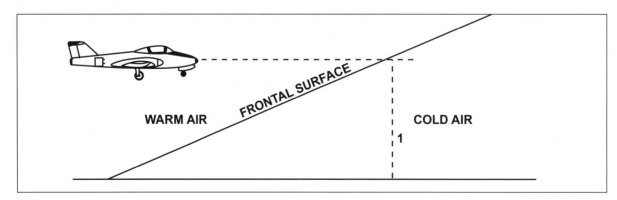

**Figure 7-9 A frontal slope of 1 in 150**

## Overrunning and Frontal Lift

9.  In Figure 7-10(a) cold air is moving to the northeast at 30 knots, warm air to the east at 20 knots and the warm front to the east at 10 knots. Figure 7-10(b) presents this situation in three dimensions. The warm air is lighter than the cold air so as it overtakes the front it will ride up over the frontal surface as shown in the cross-section in Figure 7-10(c). This is called "Overrunning." You can see that the extent of the overrunning will depend on the motion of the warm air relative to the motion of the warm front. In some cases, this will be very extensive, in others, there may not be any overrunning at all.

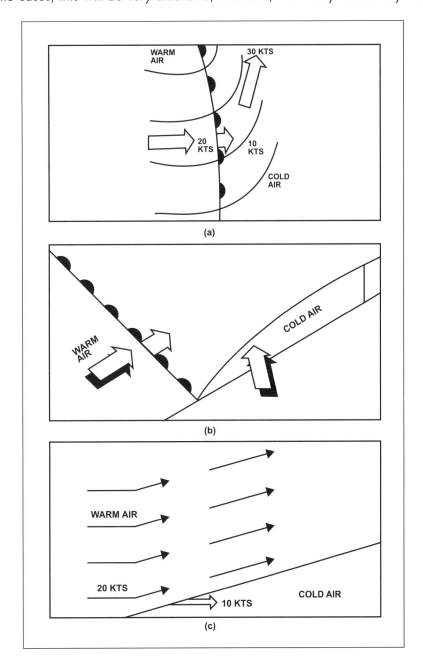

**Figure 7-10 Frontal overrunning**

10. A cold front situation is presented in Figure 7-11. Figure 7-11(a) shows cold air moving down from the north-northwest at 30 knots, the cold front is moving from the northwest at 25 knots and the warm air is moving from the west-southwest at 20 knots. This is presented in three dimensions in (b). The advancing cold air is denser and heavier than the warm air so it undercuts the warm air and forces it aloft as shown in (c). This is called "Frontal Lift." You can again see that the amount of lift is dependent on the relative motions of the front and the warm air. In some situations the lift can be very large, in others non-existent.

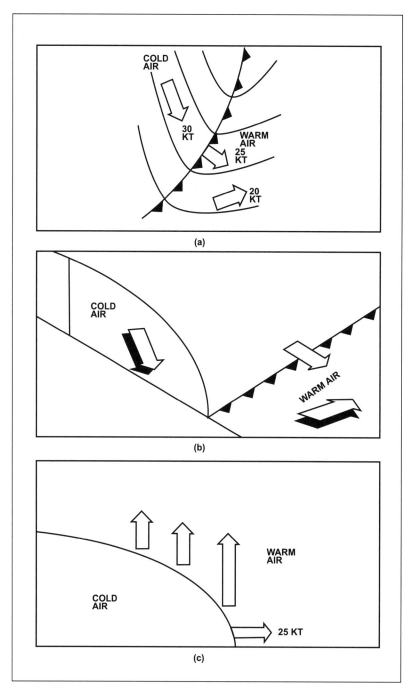

**Figure 7-11 Frontal lift at a cold front**

## Upper Fronts

11. Figure 7-12 illustrates a maritime cold front that has crossed the Rockies from the Pacific. A pool of very cold Continental Arctic air is lying stagnant across the Prairies east of the mountains. Because of the coldness and denseness of the Continental Arctic air, the front will ride over the top of it. This is called an "Upper" cold front. These upper fronts can occur anywhere that very cold air is trapped on the surface and they can be either cold or warm fronts.

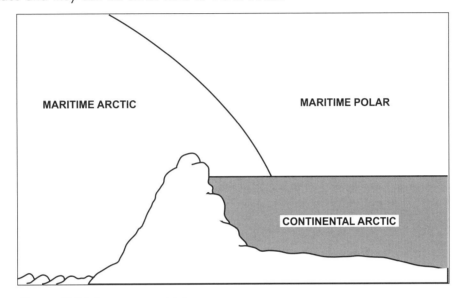

**Figure 7-12 An upper cold front**

12. Another type of upper front is shown in Figure 7-13. In this case the frontal surface is very shallow for a considerable distance and then steepens abruptly. The line along the frontal surface where it steepens is also called an upper front and this too can occur with either a cold or a warm frontal surface.

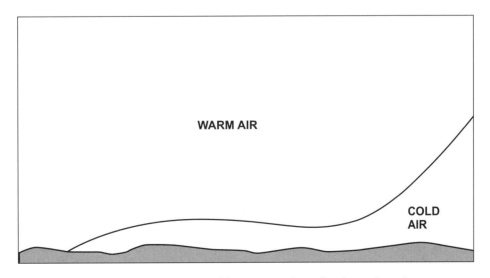

**Figure 7-13 An upper front caused by steepening of a frontal surface**

13. A final type of upper front is illustrated in Figure 7-14. You will recall that fronts exist because of a temperature difference between air masses. You should also recall that daytime heating of the atmosphere occurs from the surface upwards. There are occasions when the lower few thousand feet of the cold air can be heated sufficiently due to daytime heating and that there is no longer a sufficient temperature contrast between it and the warm air for a front to exist in this layer. At levels above this modification the front will remain. This occurs most commonly with warm fronts but can also occur occasionally with cold fronts.

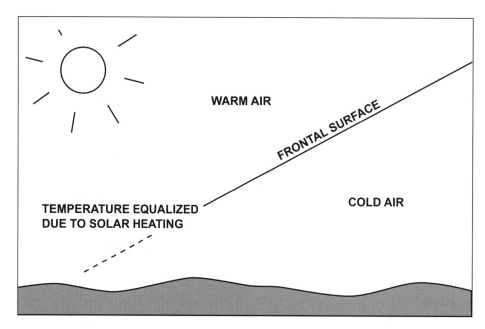

**Figure 7-14 An upper front caused by low-level air mass modification**

## Discontinuities Across Fronts

14. There are differences in the properties of adjacent air masses such as temperature, moisture, wind and visibility that are evident when you fly through a frontal surface or when a front passes over a ground station. Air mass contrast is greatest near the earth's surface, so these differences are greatest at low levels and much less by 15,000 or 20,000 feet. Because of the slope of the frontal surface, you will encounter it when you are flying at some distance from the surface position of the front. This distance will depend upon your altitude and the slope of the front.

### Temperature

15. At the earth's surface the passage of a front is characterized by a noticeable change in temperature. The amount and rate of change are indications of the front's intensity. Abrupt and large temperature changes indicate strong fronts and gradual and small changes indicate weak fronts. When flying through a front, you will note a more pronounced change at a low altitude than at a high altitude.

## Dew Point

16. Since cold air masses are generally drier than warm air masses, the dew points reported by observing stations in the cold air will normally be lower than those reported in the warm air.

## Pressure

17. Fronts always lie in troughs similar to those shown in Figure 7-15. At the earth's surface the pressure will fall with the approach of a warm front and then become steady or slowly rise following the frontal passage. When a cold front passes over a station, the pressure will begin to rise. This pressure difference will show up as changes in altimeter settings as you fly through fronts.

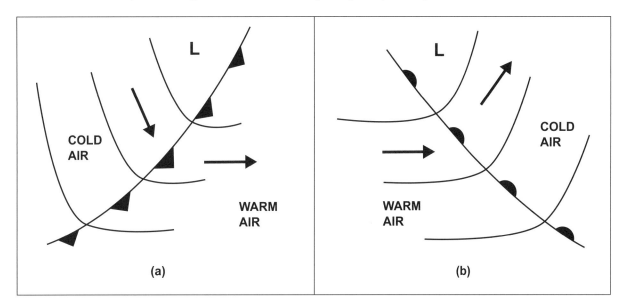

**Figure 7-15 Frontal troughs**

## Wind

18. Because fronts lie in troughs, the wind will change direction as you fly through the frontal surface. The wind speed in the cold air is frequently stronger than in the warm air. The amount of wind shift and the difference in speed is less at higher altitudes than at lower altitudes. The alteration of aircraft heading to maintain a constant track will always be to the right when flying through a cold or warm front. This alteration is the same whether you are flying from warm air to cold air or from cold air to warm air.

## Visibility

19. You will frequently find when you fly through a frontal surface into another air mass that the visibility changes markedly. Maritime Tropical air in particular is normally very hazy whereas Continental Arctic air is generally crystal clear.

# Frontal Waves and Occlusions

## Stable Waves

20. Wave-like disturbances form on fronts with areas of thick cloud and precipitation associated with them. These disturbances normally form on quasi-stationary fronts. In the initial conditions in Figure 7-16(a), the front is quasi-stationary and the winds on both sides are blowing parallel to it. This is a finely balanced situation and there are many factors that can upset it. Any divergent flow at upper levels will cause a fall of pressure. Any displacement of the front itself for whatever cause will result in a localized change in pressure. Either of these will induce a small distortion in the wind flow which further displaces the frontal boundary, causing further changes of pressure and inducing a bend in the front as in Figure 7-16(b) and (c). This is known as a "Frontal Wave." The peak of the wave is called the "Wave Crest" and the warm air portion of it the "Warm Sector." A cyclonic circulation is set up and one section of the front begins to move as a warm front and the other section as a cold front. A small low forms, centred on the wave crest. This is sometimes termed a "Frontal Depression."

21. The wave may develop no further than this and move along the front with its associated weather as in Figure 7-16(d), at a speed dependent on the upper air flow. The speed will be typically around 15 to 20 knots but in extreme cases it can be up to 40 to 50 knots. This type of wave is called a "Stable Wave." It may last for two or three days moving along the front before dissipating. Figure 7-17 is a three-dimensional illustration of a frontal wave.

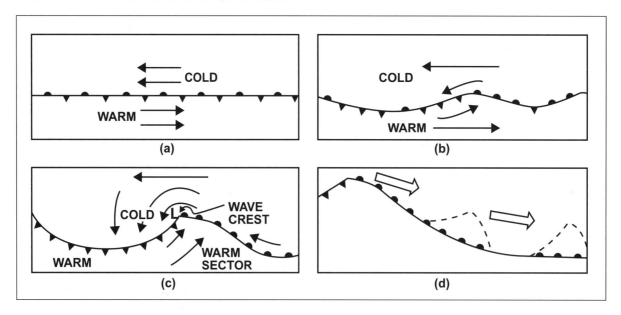

**Figure 7-16 Formation of a frontal wave**

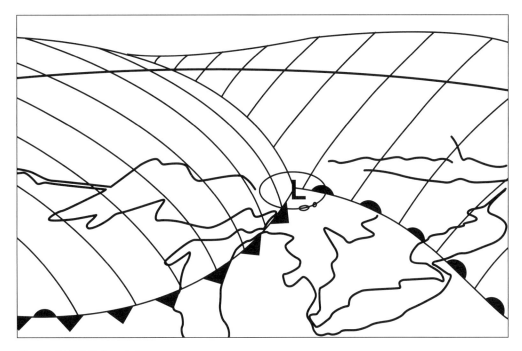

**Figure 7-17 A frontal wave**

## Unstable Waves and Occlusions

22. Under certain atmospheric conditions, the low pressure area will continue to deepen and as it does so the wind speed will increase. The wind behind the cold front increases more than elsewhere and the cold front speeds up. It begins to catch up to the warm front and the two fronts occlude (close together). This process is occurring in Figures 7-18(a) and (b). The result is called an "Occlusion" and it is the time of maximum intensity of the low.

23. The cold air mass to the northwest of the low behind the cold front is moving down from the north and will usually be colder and denser than the cool air mass ahead of the warm front which is moving up from the south. As the cold front, with its denser air, advances on the warm front, it undercuts it and forces it aloft. This forms a trough of warm air aloft called a "Trowal" shown in Figure 7-18(c) and (e). The process continues with the trowal being forced higher and higher and the warm sector becoming smaller. Finally, the warm sector disappears, the trowal is lifted to great heights and drifts away and the low is left as a cyclonically swirling mass of cold air, Figure 7-18(d). During the occlusion process, the speed of the low along the front decreases until it becomes almost stationary and the low slowly fills until it gradually disappears. The whole process, from the initial formation of the wave to the final dissipation of the low, takes several days. At its maximum size the low could cover half of Canada.

24. In Canada, the trowal is normally marked on weather maps, but in most other countries, the occluded front is marked. This is only a matter of convention; the weather will be the same in either case. The situation described has been a "cold front type occlusion" because the cold air from the north was colder than the air ahead of the warm front. On rare occasions, the air ahead of the warm front may be the coldest and in this case, a "warm front type occlusion" would form.

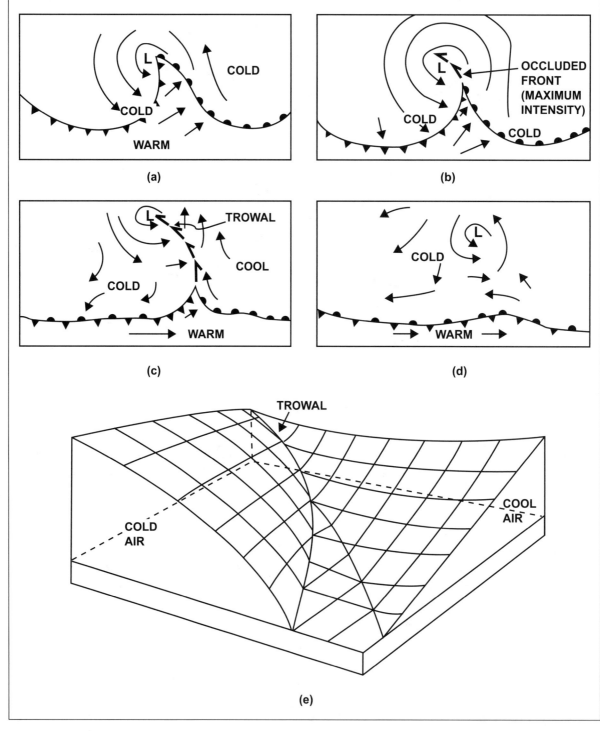

**Figure 7-18 The occlusion process**

25. All of the frontal symbols that are used on weather charts are presented in Figure 7-19.

| SYMBOL IN COLOUR | SYMBOL IN BLACK AND WHITE | FRONTAL PHENOMENON |
|---|---|---|
| continuous blue line |  | cold front |
| continuous red line |  | warm front |
| alternately red and blue line |  | stationary front |
| continuous purple line |  | occluded front |
|  blue / red |  | trowal (trough of warm air aloft) |
| broken blue line / broken red line | 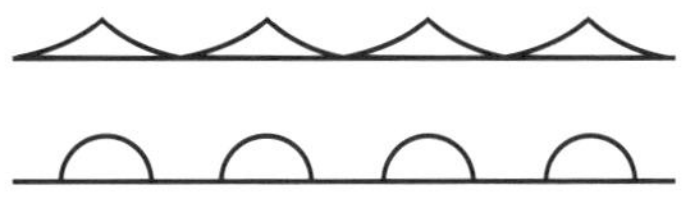 | upper cold front / upper warm front |
| blue or red | 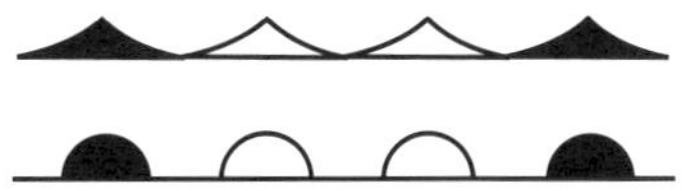 | upper cold or upper warm front or upper warm front becoming surface |
| blue or red |  | cold frontogenesis / warm |
| blue or red or purple |  | cold / warm frontolysis / occluded |
| purple |  | line squall |

Figure 7-19 Representation of fronts on weather charts

## Summary - Chapter 7

- A cold air mass lies in a huge inverted saucer shape above the earth's surface surrounded by warmer air. The cold air may reach the tropopause.

- If there is a third or fourth air mass, they will lie on top of and surround the others with the lightest and warmest air on top.

- The edge of the cold air on the surface is a front and the sloping side is a frontal surface.

- A frontal inversion is associated with the mixing zone and the frontal surface.

- Fronts move with the speed of the cold air perpendicular to the front.

- The warm air motion does not affect the frontal motion.

- Frontal surfaces slope up at an angle of about one degree with a warm front sloping a little less and a cold front a little steeper.

- Warm air overruns cold air at a warm front.

- Cold air undercuts warm air and causes frontal lift at a cold front.

- Upper fronts occur if a frontal system rides over a stagnant mass of very cold air, if a shallow frontal surface suddenly steepens or if the temperature contrast near the surface of the two air masses decreases.

- There are discontinuities of temperature, moisture, pressure, wind and visibility through a frontal surface.

- Stable waves form on quasi-stationary portions of a front and travel along it.

- A wave consists of a small cold front, warm front, a wave crest, a warm sector and a frontal low.

- In an unstable wave the low intensifies, the wave occludes and a trowal forms. The low slows down and becomes stationary then slowly fills as a cyclonically whirling mass of cold air as the trowal and warm sector disappear.

# Chapter

# 8

You have control over most of the factors involved with flying, but you have no control over the weather. Good or bad, you must adapt your flying to it.

# CHAPTER 8

## THE FORMATION OF CLOUDS AND PRECIPITATION

1.  Clouds and precipitation can obscure surface features and other aircraft. They can envelope high ground and may be associated with severe turbulence and ice formation. Modern instrumentation and aviation aids help alleviate the adverse effects of cloud and precipitation in aircraft operations but they cannot eliminate them entirely. This chapter will describe the formation and dissipation of clouds, indicate their names and how they are classified and provide some information on the characteristics of the various cloud types affecting aviation.

2.  Clouds form when water vapour condenses into liquid water droplets or sublimates into ice crystals. It was pointed out in Chapter 1 that the amount of water vapour in the atmosphere varies. It tends to be greatest near source regions such as oceans, lakes and vegetation. Saturated warm air holds much more water vapour than saturated cold air so cooling saturated warm air will result in more water vapour condensing to water droplets than will the cooling of saturated cold air. Condensation nuclei are always present in sufficient quantity for condensation to occur; however, if they are particularly abundant, condensation can occur at less than 100% relative humidity. The stability or instability of the air will determine whether the cloud forms in horizontal sheets (stratiform) or builds up in towers (cumuloform). The extent and type of cloud and precipitation produced will depend on the amount of water vapour available, the abundance of nuclei, stability of the air and the amount of cooling that the air undergoes.

3.  The major cooling mechanism that causes condensation is adiabatic expansion. There are five processes which result in the ascent of air and adiabatic cooling and they are:

    a.  Convection;
    b.  Mechanical turbulence;
    c.  Frontal lift;
    d.  Orographic lift; and
    e.  Convergence.

    Two other causes of cooling are, first, the evaporation of rain falling from higher clouds and, second, the advection over a surface colder than the air. These processes were all described in Chapter 3 in relation to atmospheric cooling. How they produce clouds will now be described.

**8**
CHAPTER

## Clouds Formed by Convection

4.   Convection was described in Chapter 4, "STABLE AND UNSTABLE AIR." It consists of air rising rapidly in shafts or bubbles and it occurs in unstable air. A layer of air can be made unstable when it is heated from below either by moving over a portion of the earth's surface warmer than itself or by the earth's surface being heated by the sun.

### Convection due to Solar Heating

5.   As the sun heats a land surface, the lower layers of the atmosphere warm and become unstable. Figure 8-1 illustrates a typical temperature structure of the lower 5,000 feet of the atmosphere. The evironmental lapse rate (ELR) is drawn as a solid black line. The inversion that develops overnight in the lower levels due to night-time cooling is indicated by the broken line in the lower 800 feet with a surface temperature of 16.5°C. The development of instability as this layer is warmed during the daytime is indicated by the heavy solid line in the lower levels where the surface temperature has warmed to 22°C. The lowest 100 feet or so of this layer has a super adiabatic lapse rate, the rest of it has a dry adiabatic lapse rate. The dew point remains at 16°C throughout the heating because moisture has been neither added nor subtracted from the air. This type of daytime heating occurs only over a land surface and does not occur over water and the extent that it occurs over land depends on the type of surface. Rocky land and dry black loam heat markedly. Evaporation from growing vegetation or moist soil, on the other hand, will reduce the amount of heating.

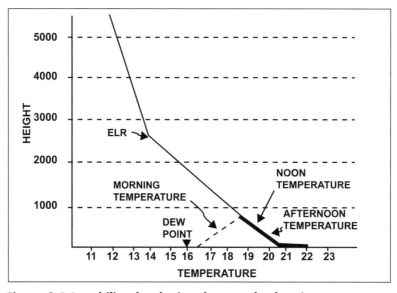

**Figure 8-1 Instability developing due to solar heating**

6.   The initiation of upward movement to a parcel of air could occur because of an eddy of wind, or because a portion of the earth's surface gets a little warmer than the rest. Once it has started upwards, the parcel will cool at the DALR (3°C/1,000 feet). Starting at 22°C the parcel will have reached 19°C at 1,000 feet where the surrounding air is 18°C so it will continue to rise. At 2,000 feet it would be 16°C and the air around it 15.5°C. As long as it is warmer than the surrounding air it will rise. This is illustrated in Figure 8-2.

7.  The moisture content of this rising bubble is represented by the dew point, which is 16°C. As the bubble rises, the dew point changes only slightly. By 2,000 feet the temperature has cooled to the dew point and cloud will begin to form. As condensation occurs, the latent heat of vaporization is released to the rising bubble of air and reduces its rate of cooling from the DALR to the SALR.

8.  As long as the parcel remains warmer than the environment the cloud will continue to grow. At 4,000 feet, both the parcel and the environment are at 13°C. If the parcel rises further it will become cooler than the environment and will sink. The top of the cloud occurs at this level.

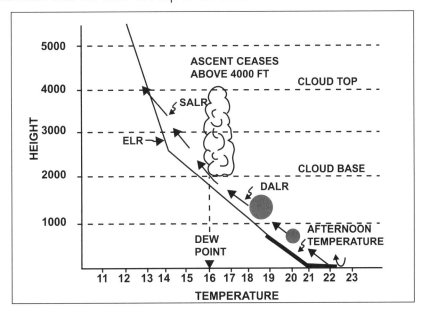

Figure 8-2 Convective cloud formed by surface heating

## Convection due to Advection

9.  Convection can be caused by the movement of air over a surface warmer than itself. One example of this is the movement of air out of the Arctic over the Pacific. The same change in the low-level ELR occurs as with solar heating. The warming from below in this type of situation does not depend on daytime heating and so convection will continue day or night as long as the flow remains the same. Another example of advective cooling is a flow of air with below-freezing temperatures over open lakes. The water temperatures will be above freezing and this heat source will develop convection (Figure 8-3).

Figure 8-3 Convection due to advection

## Height of Cloud Bases

10. In the left hand portion of Figure 8-4, convection is occurring and cloud is forming in air with a surface temperature of 20°C and a dew point of 8°C. By 4,000 feet the temperature of the rising bubble will have reached 8°C so this will be the base of the cloud. On the right hand side of the diagram the air has warmed to 29°C due to daytime heating. There has been no change in the water vapour content of the air so the dew point has remained at 8°C. The convection must now reach 7,000 feet before the air is cooled to the dew point so the base of the cloud has lifted to 7,000 feet.

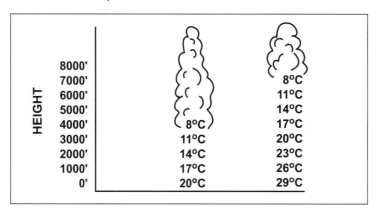

**Figure 8-4 Cloud base lifting due to an increase of the surface temperature**

11. A change in the moisture content of the air as reflected by a change in the dew point will also raise or lower the cloud base depending on whether the dew point has decreased or increased. The closer the dew point is to the temperature, the lower the cloud base.

12. The lifting of the cloud base over land as daytime heating progresses is a normal occurrence. The same thing can occur with advective heating over water but not to the same extent. As the air travels for a longer distance over the water and becomes warmer, the cloud base may rise. The dew point will also increase, however, because of the evaporation from the water, so the rise of the cloud base will not be very great.

13. The discussion so far has assumed that the dew point remains unchanged as the air parcel rises and that the rising air does not mix with the air surrounding it. In actual fact, the dew point does lower slightly with ascent and some mixing does occur. The net result is that the base of the cloud will be somewhat higher than indicated by the method described. Clouds formed in this manner are of the cumuloform type. They are frequently based around 3,000 to 5,000 feet. However, over the Prairies with very dry air and intense surface heating they can occasionally be based as high as 10,000 to 12,000 feet.

## Height of Cloud Tops and Cloud Type

14. The cloud will grow in height as long as the rising air within it remains warmer than the air surrounding it. The heating and cooling influence of the earth's surface affects only the lower few thousand feet of the atmosphere. How high convective cloud will grow depends on the instability of the air above this level. If the air is stable above it, only fair weather cumulus (CU) will form. If the air is unsta-

ble through the mid levels of the troposphere, towering cumulus (TCU) will form; and if it is unstable up to the tropopause, cumulonimbus (CB) will form. Fair weather cumulus is typically about 1,000 feet thick, towering cumulus about 10,000 feet and cumulonimbus from 20,000 to over 40,000 feet thick.

15. During convection, the air rising in the convective cells is counterbalanced by air slowly descending all around the cells. The descending air is heated adiabatically and remains cloud-free. For this reason cloud formed by convection through surface heating cannot become overcast but will always have breaks in it.

**Figure 8-5 Fair weather cumulus (CU)**

**Figure 8-6 Towering cumulus (TCU)**

**Figure 8-7 Cumulonimbus (CB)**

## Clouds Formed by Mechanical Turbulence

16. Mechanical turbulence refers to an eddying motion of the air caused by friction between the air and the ground as the air flows over the earth's surface. The intensity of the mechanical turbulence and the height to which it will extend depend on the roughness of the underlying surface, the strength of the wind and the instability of the air (Figure 8-8).

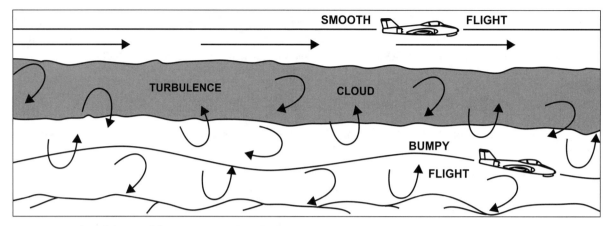

**Figure 8-8 Cloud formed by mechanical turbulence**

17. If the wind is of moderate strength, mechanical turbulence will cause air to mix in the lower levels. Figure 8-9 shows the result of mixing. The initial ELR is drawn as a broken line. In the general turmoil in the layer, particles of air aloft will be brought down to the surface and will warm dry adiabatically. Those from the top of the layer will follow path A. Similarly particles from the bottom of the layer will cool dry adiabatically as they rise to the top along path B. With mixing the actual temperature within the layer will be the average of these two extremes and is marked as a heavy solid line. The result is a dry adiabatic lapse rate topped by an inversion. If the rising particles are cooled to their dew point, cloud will form and further lift will be moist adiabatic. There will always be an inversion at the top of the mixed layer. Flight in and above the inversion will be smooth whereas it will be rough below it.

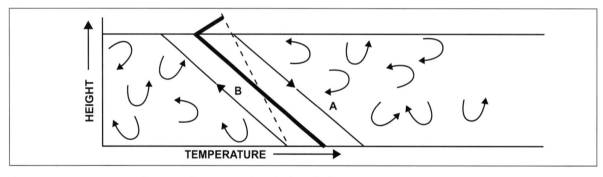

**Figure 8-9 Lapse rate changes due to mechanical turbulence**

18. The cloud formed in this manner will frequently be stratocumulus (SC). It will have an undulating base which is lower in the rising eddies than in the sinking eddies. Over land, the bases become lower as the land cools at night and the temperature–dew point spread decreases. On the other hand, increasing land temperatures during the day tend to bring about larger temperature–dew point spreads and lifting bases which, in turn, can lead towards the dispersal of the cloud or its transformation to cumulus. Over the open sea there is little diurnal variation in these clouds.

19. Because there is an inversion at the cloud top, it tends to be very flat. Occasionally, convective currents develop, embedded within the stratocumulus and these will be seen as cumulus tops protruding through the stratocumulus. Stratocumulus is frequently based between 1,500 feet and 5,000 feet and is generally about 3,000 feet thick. These heights, however, depend upon the factors indicated in paragraph 16. If the air is very moist, stratus cloud based at a few hundred feet may form instead of stratocumulus.

Photo courtesy of NASA Langley Research Center

**Figure 8-10 Stratocumulus (SC)**

## Clouds Formed by Frontal Lift

20. In considering frontal structures in Chapter 7 you learned that it was the warm air that was lifted at a front and that the amount of ascent depended on the relative motions of the frontal surfaces and the warm air. There are three factors that need to be considered in relation to the formation of clouds at fronts.

    a. THE AMOUNT AND RATE OF ASCENT OF THE WARM AIR – This, in turn, is dependent on the slope of the front and the degree of undercutting (cold front) or overrunning (warm front). Ascent is invariably greatest near frontal waves, particularly when the frontal low is deepening. In this area, lift from convergence combines with frontal lift to increase the amount of ascent of the air. The least amount of ascent occurs near the quasi-stationary portion of fronts. The rate of ascent of the air, even with active fronts, is much less than the rate of ascent that occurs with convection. With frontal lift, air is rising in the order of hundreds of feet per hour compared to hundreds or thousands of feet per minute in convective cells.

    b. THE STABILITY OR INSTABILITY OF THE WARM AIR MASS – If the air is moist and stable, a layered cloud type will form, whereas, if it is unstable, convective cloud will form. Air with potential instability is made unstable when lifted and, in this case, convective cloud will form and be embedded within layered cloud.

    c. THE MOISTURE CONTENT OF THE WARM AIR MASS – If the air is moist throughout its depth, then thick cloud will form throughout, but, if it is moist only at mid levels, then only there will cloud form. Since maritime air masses contain more water vapour than continental air masses, and since the warmer an air mass is, the more water vapour it can hold, hot, moist air masses develop far more weather than cold, dry air masses.

21. From the foregoing, it should be evident that your encounters with fronts will vary accordingly from those with clear skies to those with hazards such as hail, turbulence, low cloud and poor visibility.

## Warm Fronts

22. The amount and type of warm front cloud and precipitation will depend upon the character of the warm air and the extent that it is lifted. Because of the very shallow slope of the warm front, the area covered by cloud can be extensive, stretching possibly 1,000 miles ahead of the surface position of the front. Figure 8-11(a) and (b) illustrate weather at a warm front with stable and potentially unstable moist warm air. The situation with drier stable air is shown in Figure 8-11(c).

23. In approaching an active warm front from the cold air side, there is a typical sequence of clouds that will be encountered. The first sign of the front, possibly 1,000 miles away, will be cirrus cloud (CI) (Figure 8-12) which will thicken to cirrostratus (CS) (Figure 8-13). Altostratus (AS) (Figure 8-14) or altocumulus (AC) (Figure 8-15) will then be evident and may blend with the cirrostratus or exist as a separate layer below. If the warm air is unstable in mid levels, convective cloud will be embedded in the mid-cloud layer. The middle cloud will expand to thick altostratus and finally to nimbostratus (NS) (Figure 8-16) with precipitation. Lower stratus cloud (ST) (Figure 8-17) may form in the precipitation. The cloud sequence depends on the stability and moisture content at various heights in the warm air so that at times there may just be middle and high cloud, at other times embedded cumulonimbus based at 8,000 or 9,000 feet and, still at other times, cloud down to 100 feet.

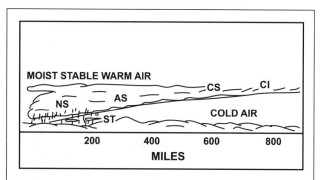

(a) **Warm front weather with moist stable warm air.**

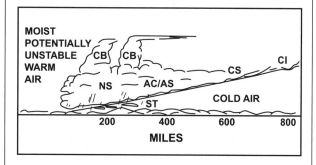

(b) **Warm front weather with moist potentially unstable warm air.**

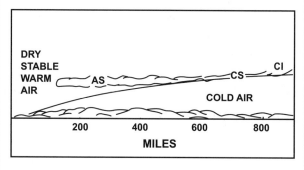

(c) **Warm front weather with dry stable warm air.**

**Figure 8-11 Warm front weather**

**Figure 8-12 Cirrus (CI)**

**Figure 8-13 Cirrostratus (CS)**

**Figure 8-14 Altostratus (AS)**

**Figure 8-15 Altocumulus (AC)**

**Figure 8-16 Nimbostratus (NS)**

**Figure 8-17 Stratus (ST)**

## Cold Fronts

24. Considering the factors that determine frontal weather, if the warm air mass is moist and unstable, and particularly if the cold front is moving fast, the warm air mass will be lifted vigorously by the wedge of cold air, and heap clouds and showery precipitation will occur. The ascent at the front may cause a potentially unstable air mass to become unstable, in which case convective type cloud, such as towering cumulus and cumulonimbus can develop and will be embedded in layer cloud. Figure 8-18(a) provides an example of a fast moving cold front.

25. A slow moving cold front will have a more shallow slope so that the rate of warm air ascent will not be as great as with a fast moving front. If the warm air is moist and stable, the cloud will be stratiform and will be more extensive than with a fast moving front. There will be less tendency for the warm air mass to become unstable due to lift. However, if it is potentially unstable, embedded convective cloud will form. Figure 8-18(b) provides an example of a slow moving cold front with moist, potentially unstable, warm air. The extent of the cloud cover with either a fast or a slow moving cold front is much less than that of a warm front.

## Squall Lines

26. Under certain atmospheric conditions, a squall line composed of thunderstorms may develop 50 to 300 miles ahead of a fast moving cold front. If a squall line does develop, there is a tendency for the cold front itself to become inactive as shown in Figure 8-18(c).

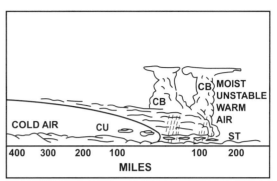

(a) A fast moving cold front with moist unstable warm air.

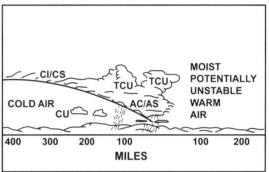

(b) A slow moving front with moist potentially unstablewarm air.

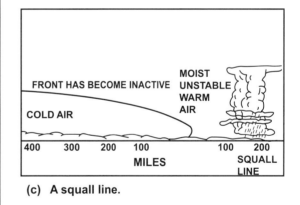

(c) A squall line.

**Figure 8-18 Cold front weather**

## Trowals

27. The weather that occurs with trowals varies considerably, but in general it is a combination of cold and warm frontal conditions with the added factor of the trowal sloping up from the crest of the wave to the end of the trowal as shown in Figure 8-19.

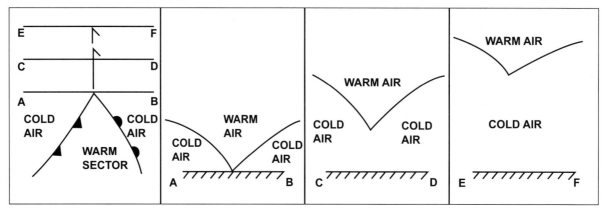

**Figure 8-19 The slope of a trowal**

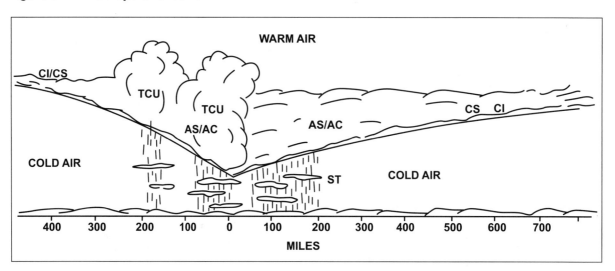

**Figure 8-20 Weather at a trowal**

28. Figure 8-20 illustrates the weather at a trowal. The warm air will likely have the same moisture characteristics at both the cold and warm front parts of the trowal. However, the more abrupt lift at the cold front portion may develop instability that is not present over the warm frontal surface.

29. As shown in Figure 8-19, the base of the trowal rises as it extends away from the wave crest. As the trowal base gets higher, the air within it becomes colder and can hold less water vapour. The potential for dense cloud and hazardous weather therefore is greatest near the wave crest and becomes progressively less with distance from the crest.

## Clouds Formed by the Evaporation of Precipitation

30. Cooling by evaporation takes place when rain falls into cool air. Heat from the air is used to evaporate the rain so the air is made cooler, and moisture is added by the evaporation of the raindrops. If the air is stable so that there is little mixing, saturation and condensation can occur and cloud can form. The cloud formed will be a layered type called "Stratus."

31. Stratus is one of the most important clouds for pilots because it forms at very low altitudes, typically around 200 or 300 feet and so presents a hazard for take-off, landing or low flying. It will form in any of the situations that have been described where precipitation has fallen long enough to saturate the air but it is most common near an active warm front.

32. Figure 8-21 illustrates the cloud shields associated with the frontal systems of an active frontal wave. This type of depiction is frequently used on weather charts to outline areas of cloud.

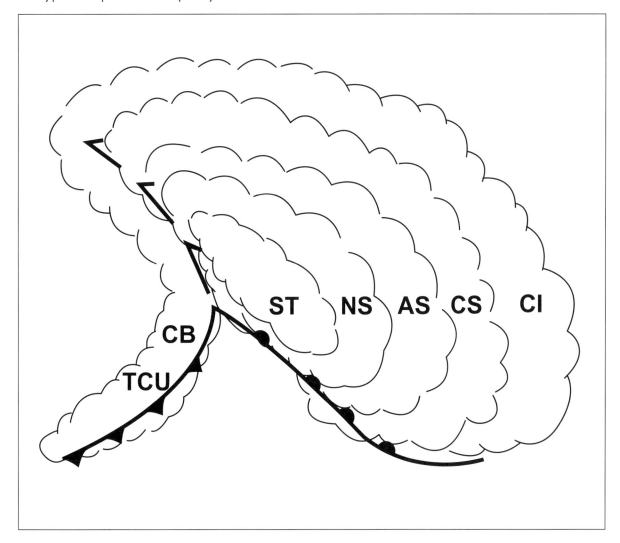

**Figure 8-21 A frontal wave cloud shield**

## Freezing Level

33. The variation in the height of the freezing level at fronts or trowals is of particular importance because of its relationship to aircraft icing. Figure 8-22 shows a frontal wave on the left and a cross-section through this wave on the right. The height of the freezing level is indicated by the 0°C isotherm marked as a broken line. Note that the freezing level is highest in the warm air and drops down through the frontal zone to a lower level in the cold air. The change in altitude can be several thousand feet.

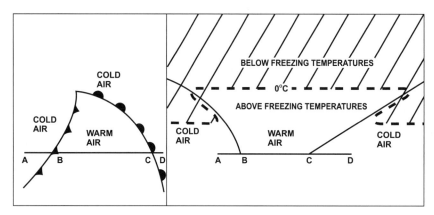

**Figure 8-22 The change in the freezing level through a front**

## Clouds Formed by Orographic Lift

34. The extent and rate of ascent of air undergoing orographic lift depends on the slope and height of the terrain and the strength of the wind component that produces the upslope flow. The rate of ascent can vary greatly from an almost imperceptible amount, if the air is slowly moving up a gradually sloping plain, to hundreds or even thousands of feet per minute, if it is moving rapidly up a mountain face. The extent of the cloud that forms depends on the moisture in the air and the type of cloud, and on the stability of the air. The air descending on the downwind side of a slope will be compressed and heated and this will cause cloud in this area to dissipate.

35. Examples of cloud formed by orographic lift are shown in Figure 8-23. If the air is dry, little if any cloud will form. If the air is moist and stable, layer cloud will form. There are frequently stratified moist layers at mid and high levels in an air mass. When these air masses undergo orographic lift, clouds do not necessarily hug the slopes, but will form in these moist layers. The vertical motion caused by the topography decreases with altitude so that at higher levels the flow has become smooth as illustrated by the series of arrows at the top of the diagrams (a) (b) and (c).

36. If the air is moist and unstable or potentially unstable, convective cloud will form. For example, cumulonimbus clouds will frequently form on the windward slopes of the coastal mountain ranges and persist day and night until the nature of the air mass changes or the direction of the flow changes.

37. Figure 8-23(d) illustrates a gradual upslope of stable air that is moist in the lower levels and which is being carried across the Canadian Prairies by an easterly wind. Such a situation as this is particularly important because it can develop a widespread area of low stratus cloud that can be particularly hazardous to aviation.

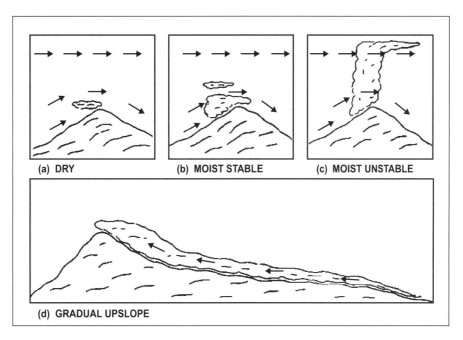

**Figure 8-23 Orographic lift**

## Clouds Formed by Convergence

38. The cloud and rain associated with lows and troughs and the fine weather associated with ridges and highs are due to slow air mass ascent and descent caused by convergence and divergence.

39. Convergence takes place in low pressure areas and troughs because of the cross-isobar flow into the low pressure. Divergence occurs in high pressure areas and ridges because of the cross-isobar flow out of the high pressure. The convergent flow into lows is greatly increased if the low is deepening and the divergent flow out of highs experiences a like increase, if the high is building.

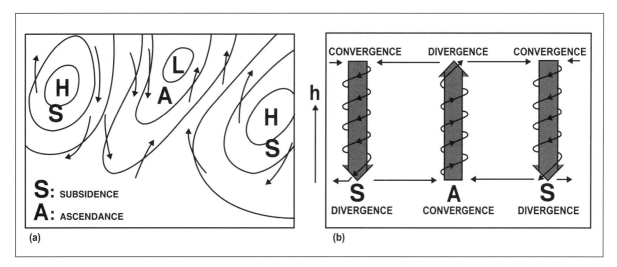

**Figure 8-24 Convergence and divergence**

40. Convergence and divergence associated with pressure systems are shown in Figure 8-24(a,b). This situation leads to ascent with a compensatory divergent flow at high levels over lows and troughs and descent with a compensatory convergent flow over highs and ridges. Figure 8-24(b) is a vertical cross-section through the two highs and the low and shows the air spiralling down in an anti-cyclonic flow over the highs and spiralling up in a cyclonic flow over the low. The vertical motion may extend only a few thousand feet into the atmosphere in weak systems, or throughout the troposphere in stronger systems.

41. The extent and type of clouds associated with convergence can vary greatly and depend on the moisture content of the air and its stability as well as on the horizontal and vertical extent of the convergent area. A potentially unstable and moist air mass can develop instability through ascent so at times convective cloud can be embedded in layer cloud.

42. The general rate of ascent in convergent areas is very slow and is in the order of 100 to 500 feet per hour. If convection develops in this ascending air, the rate of ascent in the convective cells would increase to something in the order of hundreds to thousands of feet per minute.

43. Large low pressure areas can cover half the continent with cloud due to convergence that can reach from very low levels up to the tropopause. On the other hand, minor troughs may move through an area with just a narrow belt of convective or stratus cloud.

## Clouds formed by Advection over a Colder Surface

44. When air moves over a surface colder than itself, the lowest layers are cooled by conduction and a surface-based inversion forms. When the air is cooled to its dew point, fog will form. If the wind is sufficiently strong to cause mechanical turbulence, a shallow layer of air with a dry adiabatic lapse rate develops at the surface with an inversion above it as explained in paragraph 8-17. The fog will then lift to a layer of low stratus cloud lying just under the inversion. If the cooling is particularly intense and the air very moist, the cloud base will be extremely low, and may remain as fog even with very strong winds.

45. Advective cooling is particularly prevalent over water and along coastal areas. It can cause extensive regions of low cloud and, because of this, it is of major concern to aviation. The cloud bases are frequently around 100 to 300 feet above the surface with the cloud less than 1,000 feet thick. Over water, the cloud will persist day or night as long as the airflow remains the same. Over land it will tend to dissipate due to daytime heating, but reform with nighttime cooling.

# Precipitation

46. The formation of clouds by condensation or sublimation of water vapour has been described. Whether or not precipitation will fall from the clouds depends on other factors.

47. Some idea of the relative sizes of cloud droplets, drizzle drops and raindrops can be gained from Figure 8-25. Small cloud droplets need only very weak vertical currents to keep them aloft. Drizzle droplets have a clearly visible downward velocity so that a light vertical current is needed to keep them up. Raindrops have a considerable downward velocity and strong vertical currents are needed to keep them aloft. The size of the drops or droplets varies depending upon the situation in which they are formed.

48. There is a definite upper limit to the size of the cloud droplets which form by condensation. A different process is required for these droplets to grow to precipitation size.

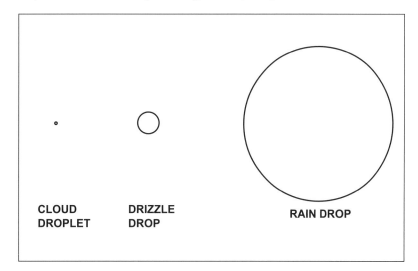

**Figure 8-25 Relative size of cloud droplets, drizzle drops and raindrops**

## Drizzle

49. Updrafts associated with stratus clouds are very weak so that the larger droplets in the cloud tend to settle earthward. As they sink they strike and coalesce with other droplets and grow. Finally they sink out of the cloud base as drizzle.

## Rain

50. Water vapour does not necessarily sublimate to ice crystals at below-freezing temperatures, but rather, can also condense to liquid super-cooled water droplets. This means that at below-freezing temperatures cloud is composed of a mixture of water droplets and ice crystals. There is an extremely strong attraction between the water droplets and the ice crystals such that the water droplets evaporate and the water vapour so produced sublimates on the crystals. In the matter of a few minutes something in the order of a million water droplets can evaporate and sublimate on one crystal. The crystals will be held

aloft in the cloud until they grow large enough that their terminal velocity exceeds the updraft velocity in the cloud. If they then fall into above-freezing temperatures they melt and coalesce with other drops and grow larger, forming raindrops. The stronger the cloud updraft, the larger the raindrops and the more intense the rain. For you as aircrew this can also be considered in reverse. The larger the drops and the heavier the rain, the stronger the updraft in the cloud and therefore the more severe the turbulence. Precipitation can form in warm climates due to coalescense alone, but in Canada, for moderate or heavy precipitation to occur, both sublimation on ice crystals and coalescense are required.

### Snow

51. If the temperatures are cold all the way to the ground, the ice crystals will not melt but will aggregate into snow flakes. Large flakes occur at temperatures just below freezing while smaller, more solid types of snow form at colder temperatures.

### Intensity of Precipitation

52. The intensity of precipitation depends on the strength of the updraft within the cloud, and the cloud's vertical thickness and water content.

53. Strong updrafts are needed to produce large raindrops. For this reason, the heaviest rainfall occurs with large convective types of cloud. The vertical thickness of cloud is important for two reasons. First, the air must be warm in order to hold a lot of water vapour. This is only possible with a low cloud base. Second, ice crystals are required to produce precipitation. The cloud top, therefore, must extend above the freezing level. These two statements imply that a thick cloud is necessary for heavy rain. Snow presents a somewhat different situation so that if other factors are suitable, heavy snow can fall out of a cloud 2,000 or 3,000 feet thick. The water content of the cloud is related to both the temperature of the cloud, as just explained, and the moisture source available for evaporation of water into the air. Tropical air that has originated over the oceans produces the heaviest precipitation. It is in this air mass that monsoons and hurricanes, with their devastating floods, occur.

### Showery, and Intermittent and Continuous Precipitation

54. Precipitation is described as Showers, Intermittent Precipitation or Continuous Precipitation:

   a. SHOWERS are of short duration, beginning and ending abruptly and with a noticeable brightening of the sky between them. They fall from cumuloform cloud, although these may be embedded in layer cloud.

   b. INTERMITTENT PRECIPITATION is not showery in character but has stopped and recommenced at least once during an hour.

   c. CONTINUOUS PRECIPITATION is not showery in character but continues without a break for at least an hour. Continuous and intermittent precipitation fall from layer clouds.

# Classification of Clouds

55. Clouds are classified into four families: High Clouds, Middle Clouds, Low Clouds and Clouds of Vertical Development. Each of the first three families is subdivided according to whether they billow up in towers (cumulus type) or lay flat in horizontal sheets (stratus type). In addition to these subdivisions, the word "Nimbus" is added to the names of clouds that normally produce precipitation.

| Group | High Clouds base 20 000 feet to 40 000 feet | Middle Clouds base 6 500 feet to 20 000 feet | Low Clouds base surface to 6 500 feet | Clouds of Vertical development base 1 600 feet up |
|---|---|---|---|---|
| Type | Cirrus (CI)<br>Cirrostatus (CS)<br>Cirrocumulus (CC) | Altostratus (AS)<br>Altocumulus (AC)<br>Altocumulus (ACC) Castellanus | Stratus (ST)<br>Nimbostratus (NS)<br>Stratocumulus (SC)<br>Stratus Fractus (SF)<br>Cumulus Fractus (CF) | Cumulus (CU)<br>Towering Cumulus (TCU)<br>Cumulonimbus (CB) |

**Figure 8-26 Classification of clouds**

## High Clouds

56. The high clouds are made up of ice crystals and are usually based above 20,000 feet.

a. CIRRUS (CI) - This cloud appears as white curly streaks across the sky.

b. CIRROSTRATUS (CS) - Unlike cirrus, this cloud appears as a whitish veil through which the sun and moon can be seen, often surrounded by a halo.

c. CIRROCUMULUS (CC) - This is a somewhat rare cloud. It appears as a white sheet with a pebbly pattern.

## Middle Clouds

57. The middle clouds are based between 6,500 and 20,000 feet. They may be collections of water droplets, ice crystals, or a combination of both.

a. ALTOSTRATUS (AS) - This is a layer cloud with no definite pattern such as rolls or waves on its undersurface. It is steely or bluish in colour and, as seen from a particular place, may cover the entire sky. Sometimes the sun or moon can be seen dimly through it but there are no halos.

b. ALTOCUMULUS (AC) - This is a layer or series of patches of rather flattened, rounded masses of cloud. The cloudlets may be arranged in groups, lines or waves and are sometimes so close that their edges join.

c.   ALTOCUMULUS CASTELLANUS (ACC) - This is similar to altocumulus but with pronounced turrets building upward. It implies considerable instability in the cloud layer and may develop into cumulonimbus.

## Low Clouds

58.   The low clouds are usually based below 6,500 feet. They include:

a.   STRATUS (ST) - A uniform layer of very low cloud that may appear in extensive sheets or irregular patches. It resembles fog except that it does not rest on the ground, although it may be very close to it. Its undersurface does not show any pattern such as waves or ripples. When it is torn by the wind it appears in fragments referred to as Stratus Fractus (SF). Drizzle or freezing drizzle may fall from it.

b.   NIMBOSTRATUS (NS) - This is the main precipitation cloud: continuous rain, snow, freezing rain, etc, may be encountered when flying in or below it. In appearance it is an extensive layer, uniformly dark in colour, that may be based from 6,500 feet to near or at the ground. Nimbostratus is often part of an extensive cloud layer that forms in the overrunning warm air ahead of a warm front.

c.   STRATOCUMULUS (SC) - This is a common and easily recognized cloud form. The bottom has a clear-cut, wavy or rolled appearance. It often appears as an extensive sheet, but sometimes there are well defined breaks between the rolls. Occasionally convective clouds are embedded in it. By itself, it gives little precipitation except in very cold weather, when it may give snow.

## Clouds of Vertical Development

59.   The convective clouds may appear as isolated clouds, or they may be embedded in layer clouds. They include:

a.   CUMULUS (CU) - These are fluffy white clouds that form in the top of convection currents. They are a common sight over land during a hot summer afternoon. Their edges are hard and clear-cut in appearance and their tops are rounded. When they appear as ragged or torn fragments, they are called Cumulus Fractus (CF).

b.   TOWERING CUMULUS (TCU) - The name aptly describes these cumulus clouds that have grown to considerable height but still have clear-cut rounded tops. Showers may fall from towering cumulus.

c.   CUMULONIMBUS (CB) - When a towering cumulus grows to a great height, perhaps to the tropopause or higher, the top loses its hard, clear-cut appearance and frays out into a widespread, white, fibrous structure, often called an anvil or thunderhead. The cloud is now a cumulonimbus or thunderstorm cloud. Heavy precipitation in the form of rain or hail showers may be seen pouring out of it. They are very dangerous for aircraft but, unless they are embedded in or obscured by other clouds, the white spreading tops can be recognized from a considerable distance.

# Cloud Characteristics Affecting Flight

60. Visibility and turbulence within different types of cloud vary a great deal depending on the stability of the air, and on whether the cloud is composed of water droplets or ice crystals. Clouds also affect radio and navaids.

    a. HIGH CLOUDS - These have little effect on flying except for moderate turbulence and limited visibility associated with dense jet stream cirrus and possibly heavy turbulence in anvil cirrus. ADF and LF radio interference occurs.

    b. MIDDLE CLOUDS - With these, visibility can vary depending on the thickness of the cloud from possibly 1/2 mile in very thin altostratus to a few feet in thick altostratus. Turbulence would normally be nil to light unless convective activity were embedded, or castellanus were developing. Rain, rain and snow mixed, or snow can be encountered in thick altostratus or well developed castellanus depending upon the height of the freezing level and the position of the aircraft in relation to it. Infrared sensors are affected.

    c. LOW CLOUDS - Visibility will be several hundred feet to a few feet in stratus and stratocumulus. Little turbulence will occur in stratus, occasionally moderate in stratocumulus. Any noticeable precipitation will be drizzle in stratus and snow in stratocumulus. The low cloud bases that can occur with stratus can cause take-off or landing difficulties. Rain, rain and snow mixed, or snow can be encountered in nimbostratus depending upon the height of the aircraft.

    d. CLOUDS OF VERTICAL DEVELOPMENT - Visibility will be from 20 to 30 feet up to 100 to 200 feet in all clouds. Turbulence will vary from light to moderate in fair weather cumulus, moderate to severe in towering cumulus and heavy to extreme in cumulonimbus. Precipitation will vary depending on the position of the aircraft in relation to the freezing level. It can include rain, rain and snow mixed, snow, or hail, and can at times be heavy. ADF, LF and infrared sensor interference occurs in cumulonimbus.

## Some Important Effects of Precipitation

61. Visibility in light rain or drizzle is somewhat restricted, however in heavy rain or drizzle it may drop to a few hundred feet. Rain or drizzle streaming across the windscreen will further restrict forward visibility to an extent that varies with aircraft type. Snow markedly reduces visibility and can lead to an almost total loss of forward vision out of the cockpit.

62. Very heavy rain falling on a runway can wet the surface sufficiently that an aircraft will hydroplane. During hydroplaning, the aircraft tires are completely separated from the actual runway surface by a thin film of water. Under these conditions the tire traction becomes almost negligible and in some cases the wheel will stop rotating entirely. The tires will provide no braking capability and will not contribute to the directional control of the aircraft resulting in loss of control.

**8** CHAPTER

63. If there is sufficient depth of wet snow on the runway it tends to pile up ahead of the tires of an aircraft on its take-off run. This can create sufficient friction so that the aircraft is unable to reach rotation speed and cannot become airborne.

64. Heavy rain ingested by the engines of a jet or turbo-prop aircraft in flight can cause power loss, or actual flame-out.

65. An encounter with hail can cause serious damage to any aircraft, but so can an encounter with rain if it is penetrated at very high speed. The impact pressure created by rain has been computed to be 18,000 pounds per square inch when flying at a speed of Mach 1.6. This pressure has been known to peel flush rivet heads out of an aircraft's leading edges, wear plexiglass down, erode fibreglass antennas and peel paint off the aircraft.

## Condensation Trails

66. An aircraft leaves a condensation trail (contrail) behind it when the moisture formed during combustion and emitted with the exhaust gases is sufficient to saturate the air, subsequently causing condensation.

67. For each pound of aircraft fuel burned, approximately 1.4 pounds of water vapour are formed and ejected with the engine exhaust gases. This increases the relative humidity in the wake of the aircraft. On the other hand, the heat generated by the engine tends to lower the relative humidity in the wake by raising the temperature. There is a gradual mixing of the exhaust with the air behind the aircraft which varies from zero immediately behind the aircraft to complete mixing a considerable distance behind.

68. In certain conditions, the net result is to increase the humidity to saturation so that cloud forms 100 or 200 feet behind the aircraft as the exhaust cools. In the case of jet aircraft, the critical conditions under which contrails form are almost the same for all types of aircraft.

69. Whether a trail will form or not depends on the temperature and relative humidity of the air surrounding the aircraft. This is illustrated in Figure 8-27. Contrails will not form to the right of the sloping 100% relative humidity line, but they will form to the left of it depending on the relative humidity. For example at 200 hectopascals (40,000 feet) contrails will form at any temperature colder than -55°C even with 0% relative humidity and at -50°C if the relative humidity is 90%.

70. Whether the contrails will be persistent or quickly evaporate depends mainly on whether the contrail particles are composed of super-cooled water droplets (see Chapter 9 for definition) or of ice crystals. If they remain as super-cooled water droplets, mixing with the surrounding air will cause them to evaporate within half a minute or so. If they have turned to ice crystals, they may persist for hours, and indeed several contrails may merge to cause an overcast of cirrus. At temperatures colder than -40°C, water droplets freeze to ice crystals within a very short time so that it is at temperatures colder than this that contrails will normally be persistent provided they will form. They may also change to ice crystals and persist if the relative humidity is very high, for example, in thin cirrus cloud.

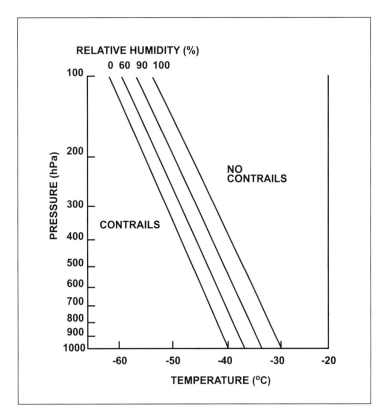

**Figure 8-27 Contrail formation**

71. Contrails make the visual detection of aircraft extremely easy so for military operations it may be important to avoid forming them. If a large number of aircraft are to rendezvous at a certain altitude, the formation of persistent contrails could cause a sky condition that would make a rendezvous at this altitude hazardous so that an altitude where contrails would not form would be preferred. For these reasons, it may be desirable to prevent the formation of contrails. Some suggestions for this follow:

   a. Fly at a level where the temperatures are warm enough that persistent contrails will not form.

   b. Fly very high in the stratosphere provided that the temperatures there are not too cold. The air will be very dry and, as shown in Figure 8-27, it takes very cold temperatures to produce contrails at great heights.

   c. Try to find a dry level. For example, a layer of air that has wisps of cirrus in it will be moist and should be avoided.

   d. Reduce your throttle setting as much as possible. The more fuel burned the more water vapour produced.

# Summary - Chapter 8

- Most clouds form as a result of cooling by adiabatic expansion.

- Clouds also form from cooling by evaporation of rain or by advection of air over a colder surface.

- CU, TCU and CB can form from surface heating (solar and advection). The type of cloud is dependent on the instability of the air above the surface layer. The bases tend to rise with daytime heating.

- Mechanical turbulence forms Stratocumulus clouds.

- Frontal activity is greatest near frontal waves. Warm front cloud is much more extensive than cold front cloud but cold front cloud tends to be more violent.

- The cloud formed during frontal lift can be layered, towering or both depending on whether the warm air is stable, unstable or potentially unstable.

- Stratus can form in precipitation from higher cloud or from advection over a cold surface.

- The freezing level is lowest in the cold air of a frontal system.

- Clouds formed by orographic lift can also be layered or towering or both depending on the type of stability of the air. The clouds dissipate downwind of the lifted area.

- Cloud can also form due to convergence. The amount and type depend on the air mass characteristics.

- Drizzle forms by coalescence of stratus cloud droplets.

- Rain forms by cloud water droplets evaporating and then sublimating on ice crystals. The ice crystals melt when they fall below the freezing level and coalesce on other drops.

- If the temperatures are below freezing, the crystals merge and fall to the earth as snow.

- The intensity of precipitation depends on the strength of the updraft and the thickness and water content of the cloud.

- Clouds are classified as high, middle, low or clouds of vertical development.

- Clouds have a variety of effects on flying that include reduced visibility, turbulence and precipitation. They also affect radio and navaids.

- Wet snow on the runway can create enough tire friction to prevent take-off.

- Heavy rain can cause hydroplaning and engine power loss or flame-out and can reduce visibility.

- Snow seriously reduces cockpit visibility.

- Hail can cause serious damage and so can rain if it is penetrated at high speed.

- Contrails will be persistent at temperatures colder than -40°C.

**8** CHAPTER

# Chapter 9

Aircraft accidents occur after a series of events place a pilot in a box from which he cannot escape. Icing is one of the events that can close the box, making an accident inevitable. Your knowledge of icing and how it affects your aircraft may prevent the last side of the box from closing.

# CHAPTER 9

# AIRCRAFT ICING

## Super-Cooled Water Droplets

1. When ice crystals are warmed to above-freezing temperatures, they melt. On the other hand, when water droplets are cooled to below freezing they will not freeze until very cold temperatures are reached. Water droplets in this state are called "super-cooled." If these droplets impact on an aircraft at below-freezing temperatures, the jar will cause them to freeze and they will coat the aircraft with ice.

## Effects of Icing on Aircraft

2. Ice on an aircraft has several very serious effects. It will disrupt the smooth laminar flow over airfoils or rotors causing a decrease of lift and an increase in the stalling speed. It will increase drag and weight. Uneven shedding of ice from propellers or rotors can cause destructive vibrations. Water can freeze around control surfaces and restrict their movement. Pitot heads and static vents can be blocked causing erroneous altimeter, airspeed and vertical speed indications. Antennas can be broken off with the resultant loss of communications and navaids. Ice can cover windscreens and block vision. Undercarriage and brakes can freeze from splash during take-off and become inoperative. Power can be lost in both jet and piston engines. Fuel consumption will increase because of increased drag and weight. Even the use of deicing/anti-icing will increase the fuel consumption because of the enormous amount of energy required to eliminate the ice.

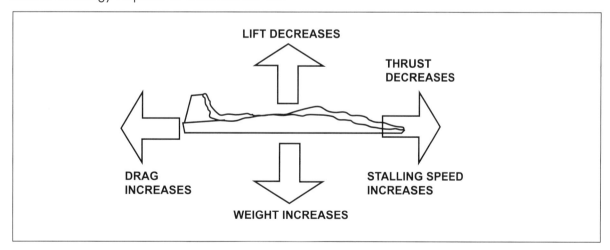

**Figure 9-1 Aircraft icing**

3. The amount of ice that collects on an aircraft depends not only on the meteorological factors involved, but just as importantly on aerodynamic factors that are related to aircraft design.

# Meterological Factors

4. Airframe icing results when super-cooled water strikes portions of the airframe that are colder than 0°C. The greater the amount of super-cooled water, the worse the icing.

## Liquid Water Content of Clouds

5. The liquid water content of a cloud is a measure of the amount of liquid water in a given amount of cloud, and is dependent on the size and the number of droplets in that given amount. The larger the liquid water content, the more serious the icing.

6. Strong vertical currents are necessary to prevent large droplets from falling out of a cloud. The strongest currents and the largest water droplets are found in convective cloud, in cloud formed by abrupt orographic lift and in lee wave cloud. * These clouds will have the largest water droplets. Clouds formed due to weakly ascending air will be composed of small droplets.

## Effects of Temperature

7. Warm air can hold more water vapour than cold air. For this reason, the amount of water droplets condensed out is greater in cloud formed in warm air masses. In convective clouds the warmer the cloud base, the greater the amount of water that may be condensed out in the clouds (because of its influence on the free water content throughout the clouds) and, thus, the more serious the icing.

8. Temperature plays another very important role both as to the size and the number of droplets in a cloud. Larger droplets begin to freeze spontaneously to ice crystals at around -10°C and, as droplets get smaller, colder temperatures are required to freeze them. By -40°C virtually all droplets will have frozen. The rate of freezing increases markedly at temperatures just below -15°C.

9. When water droplets and ice crystals exist together in a cloud, there is a tendency for the water droplets to evaporate and for the resulting water vapour to sublimate on the ice crystals. The crystals therefore grow rapidly and begin to settle downward. As they fall, they very rapidly deplete the liquid water content throughout the cloud.

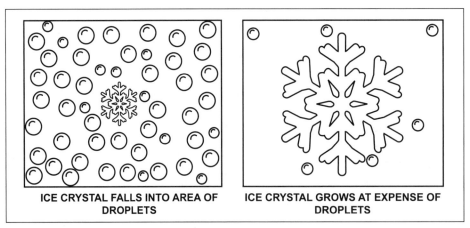

| ICE CRYSTAL FALLS INTO AREA OF DROPLETS | ICE CRYSTAL GROWS AT EXPENSE OF DROPLETS |

**Figure 9-2 Interaction of water droplets and ice crystals**

*Lee wave clouds are described in Chapter 14, "Mountain Waves."

CHAPTER 9

10. This process can cause a large and rapid fluctuation in the number of droplets in the cloud. If ice crystals begin to form, or if snow should fall into the cloud from higher cloud, the water droplet content will decrease rapidly. This growth of ice crystals at the expense of water droplets implies that the liquid water content of cloud from which snow is falling will normally be small.

11. In clouds with very strong vertical currents, such as towering cumulus and cumulonimbus, the water droplets can be carried up so fast that they will reach very high levels and very cold temperatures before freezing. In cumulonimbus, water droplets can even be found at high levels embedded in the cirrus anvil.

12. Because of all these factors, icing intensity can change rapidly with time so that one aircraft following another by only a few minutes might encounter quite different icing conditions.

13. Figure 9-3 shows a thick non-precipitating cloud with an embedded cumulonimbus. The temperature near the cloud base is above freezing and it decreases to below -40°C at the top of the cloud. At heights below the freezing level, the cloud is composed of water droplets. Between 0°C and -10°C the water droplets are super-cooled with very few, if any, ice crystals present. At temperatures warmer than -15°C, water droplets tend to predominate; at temperatures colder than -20°C, ice crystals tend to predominate. Between -10°C and -40°C the cloud composition changes from predominately water droplets to predominately ice crystals. The change occurs most rapidly at temperatures a little colder than -15°C with the larger droplets changing at warmer temperatures and the smaller droplets at very cold temperatures. Should ice crystals start falling through the cloud, they will grow and thus decrease the super-cooled water droplet content of the cloud. The strong convective currents in the cumulonimbus are shown carrying even large super-cooled droplets to very high levels.

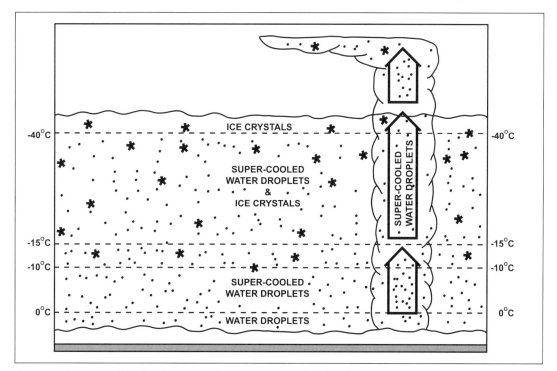

**Figure 9-3 Water droplet-ice crystal arrangement in cloud**

14. When the temperature is not too cold, the ascending air, in which a cloud forms, carries the cloud droplets up with it so that the liquid water content tends to be at maximum near the cloud top. If the temperature near the cloud top is colder than approximately -15°C, ice crystals will begin to form and the liquid water content will decrease. Because of this, icing will tend to be heaviest near cloud tops unless the tops are quite cold.

## The Freezing Process

15. When a super-cooled droplet strikes an aircraft, it begins to freeze. As it freezes it releases latent heat of fusion and this warms the remainder of the droplet to 0°C. The droplet then continues to freeze, but more slowly. The fraction of the droplet that freezes on initial impact is greatest at colder temperatures (Figure 9-4(a)). The rate of freezing after impact depends on the temperature of the aircraft skin and on the air temperature. The more closely these temperatures approach 0°C, the slower the water freezes and the more it spreads from the point of impact before freezing is completed.

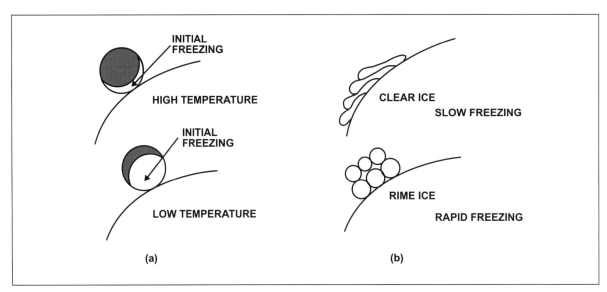

**Figure 9-4 Freezing of super-cooled droplets on impact**

16. The size of the droplets and the frequency with which they strike the aircraft are important because the character of the ice depends on whether or not each drop freezes completely before another drop strikes the same spot. If the droplets pile rapidly on each other before being completely frozen, the unfrozen parts mingle and spread out before freezing. If the droplets freeze completely before being hit by another droplet, a large amount of air is trapped causing the ice to be opaque and brittle (Figure 9-4(b)).

## Types of Ice

17. RIME - Rime is ice which is rough, milky and opaque in appearance and is formed by the almost instantaneous freezing of small super-cooled water droplets. It will usually form only on the leading edges of airfoils and tends to build forward into the air stream, forming fingers and ridges. Because of the low adhesive properties of rime, it is generally readily removed by de-icing equipment (Figure 9-4(b)).

18. CLEAR ICE - This type of ice has high adhesive and cohesive properties. Unlike rime, it can spread from the leading edges, and in severe cases may cover the whole surface of the aircraft. Its physical appearance can vary from transparent and glass-like to a very tough opaque surface. Clear ice is formed when large super-cooled water droplets collide with the airframe and freeze slowly after impact. The free water then flows back over the airfoil surface as it freezes at temperatures not far below freezing. Clear ice thus builds back from leading edges as well as forward and may develop large irregular protuberances into the air stream. It is a more serious hazard than rime ice (Figure 9-4(b)).

19. Frequently, the temperature and the range of droplet sizes are such that the ice formed is a mixture of rime and clear.

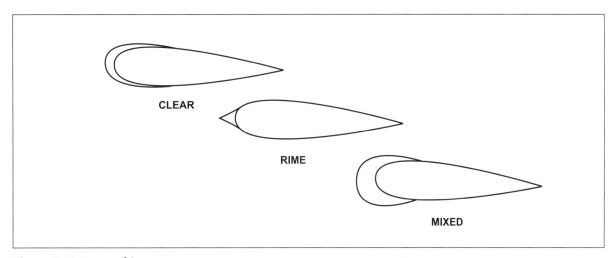

**Figure 9-5 Types of ice**

## Intensity of Icing

20. TRACE - Ice becomes perceptible. The rate of accretion is slightly greater than the rate of sublimation. It is not hazardous even though deicing/anti-icing equipment is not utilized, unless encountered for an extended period of time (over 1 hour).

21. LIGHT - The rate of accretion may create a problem if flight is prolonged in this environment (over 1 hour). Occasional use of deicing/anti-icing equipment removes/prevents accretion. Light ice does not present a problem if the deicing/anti-icing equipment is used.

22. MODERATE - The rate of accretion is such that even short encounters become potentially hazardous and the use of deicing/anti-icing equipment, or diversion, is necessary.

23. SEVERE - The rate of accretion is such that deicing/anti-icing equipment fails to reduce or control the hazard. Immediate diversion is necessary.

24. As there is no satisfactory instrument installed on aircraft for directly measuring the rate of ice accretion on an airframe, these terms must be interpreted qualitatively and measured by the effect of ice formation on the flight characteristics of the individual aircraft type. What is considered moderate icing for one aircraft type may be only light for another.

# Cloud Types and Icing

## Convective Clouds

25. Large cumulus and cumulonimbus clouds are associated with unstable air and strong updrafts. These updrafts produce large drops and may transport large amounts of super-cooled water to great heights, particularly when the clouds are developing. Cumulonimbus clouds have a cellular structure and, while some cells are growing, others may be decaying, so the composition of the cloud varies considerably at the same level. Growing cells tend to contain a high proportion of super-cooled drops, whereas ice crystals develop rapidly and tend to predominate as the cell ceases to grow. Therefore, icing is likely to be particularly severe in newly developed parts of a cumulonimbus. Fortunately, the horizontal extent of the icing is small.

26. The following are some generally accepted rules for icing in large cumulus and cumulonimbus clouds:

   a. At temperatures below -40°C the possibility of icing is small.

   b. At heights where temperatures are between -25°C and -40°C the possibility of moderate or severe icing is small except in newly developed cloud, but light icing is always possible. The type of ice will normally be rime.

   c. At heights where the temperature is between -25°C and 0°C the rate of icing is severe over a substantial depth of cloud for a wide range of cloud base temperatures but especially if they are warm. The type of ice will normally be clear.

## Layered Clouds

27. Owing to their generally weaker updrafts, icing conditions in layer clouds are, on the average, less severe than those in cumulus type clouds. Unlike cumulus type clouds, icing areas associated with layer clouds are more likely to be great in horizontal extent, but limited in the vertical.

28. Occasions of airframe icing in layer clouds are predominantly in the 0°C to -15°C temperature range and are generally light or moderate in intensity. However, trigger actions which may increase the strength of updrafts within layer clouds will also increase the severity of icing and lower the temperature at which it may be expected. For example:

   a. Frequently in winter, the air over the sea or open lakes is unstable in the lower layers only (say the first 5,000 ft) and there is a sharp inversion above. In these conditions the convective cloud tends to spread out under the "lid" of the inversion to give a layer of SC. These layers tend to contain a larger concentration of super-cooled drops and produce more severe icing than SC formed by mechanical turbulence over land.

   b. On occasions when thick layer clouds are formed by rapid mass ascent in an intensifying front, trough or low, the probability of icing is much increased. Severe icing in these conditions has been reported at temperatures as low as -20°C to -25°C.

c. The orographic effect of a range of hills is likely to increase the depth of a cloud layer and the liquid water concentration in the cloud. Thus aircraft are likely to encounter a much greater icing rate over hills.

d. The lenticular clouds sometimes formed downwind of hills or mountains can have very strong vertical currents associated with them due to mountain waves. Icing can be severe in them, and because the water droplets are large they tend to form clear ice.

## Cirrus Cloud

29. Cirrus clouds are usually composed of ice crystals, which do not constitute a serious icing hazard to aircraft. As noted before, however, water droplets may be found in thunderstorm anvils.

## Freezing Rain

30. In Figure 8-25 the relative size of rain and drizzle drops and cloud droplets were illustrated. From this it can be seen that if the larger cloud droplets tend towards clear ice, rain and drizzle drops will definitely cause clear ice.

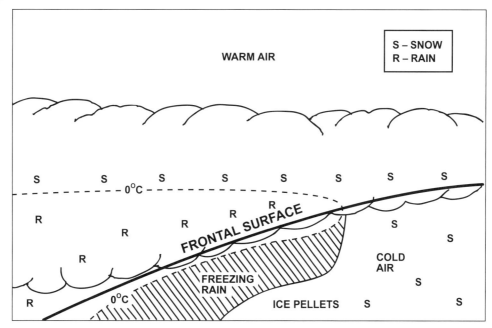

**Figure 9-6 Freezing rain at a warm front, vertical cross-section**

31. In the winter, the freezing level of the warm air mass of a frontal system is sometimes at a height above the surface while the freezing level of the cold air mass is lower and normally on the surface. This is illustrated in Figures 9-6 and 9-7.

32. In Figure 9-6 the freezing level is shown as a broken line through the middle of the cloud in the warm air and folding back under the frontal surface to lie on the ground in the cold air. On the far right of the figure, snow that has formed in the cloud falls through below-freezing temperatures all the way to

the ground. On the far left of the figure, the snow falls into the above-freezing temperatures, melts and falls to the ground as rain. In the centre of the figure, the snow falls, melts to rain as it passes into the above-freezing temperatures but then falls further into the below-freezing temperatures and becomes super-cooled. This is "Freezing Rain." If it should strike any object, it will freeze and coat the object with clear ice. If the freezing rain falls far enough through below-freezing temperatures, it will freeze into little pellets of clear ice called "Ice Pellets." A typical sequence of precipitation encountered by an aircraft flying towards the front under the frontal surface is snow, ice pellets, freezing rain and then rain. Note that if you should encounter ice pellets, it implies freezing rain above you, and if you are flying towards the front, then also, ahead of you. If you are in freezing rain, there is an above-freezing layer above you. Figure 9-7 gives a horizontal depiction of this same condition.

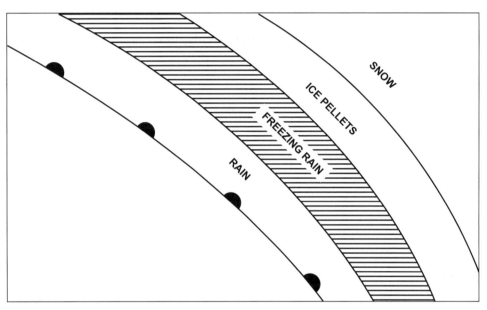

**Figure 9-7 Freezing rain at a warm front, horizontal depiction**

33. Freezing rain can also occur at trowals and at a cold front, but is not as common. Since a cold frontal surface is much steeper than a warm frontal surface, the freezing rain there will not have as large a horizontal extent.

## Freezing Drizzle

34. Unless temperatures are very cold, stratus cloud is composed of super-cooled water droplets. These may coalesce and form drizzle, which will then fall from the cloud as freezing drizzle. Since these droplets are very much larger than even the largest cloud droplets, the ice formed by them will be clear ice. The drizzle evaporates to some extent as it falls to the ground so the icing will be the most severe just near the cloud base.

35. While icing in both freezing rain and freezing drizzle will frequently be severe, the liquid water content in the precipitation is not necessarily greater than in cloud because the number of drops in a given volume may be very much less than in the same volume of cloud. For this reason, while icing in freezing precipitation should always be considered hazardous, icing in cloud should not be underestimated.

## Snow and Ice Crystals

36. Dry snow and ice crystals will not adhere to an aircraft and will not normally cause icing. If the portion of the aircraft skin that they strike is above freezing, they may melt and freeze as they flow back over below-freezing portions of the aircraft. If there is a mixture of super-cooled water droplets and snow, a rapid buildup of very rough ice can occur.

37. Wet snow that has fallen on an aircraft and is not removed can freeze hard by evaporation cooling once the aircraft is in motion. This can occur even with temperatures a little above freezing. Slush or water from the runway splashed on the undercarriage or wheels can freeze in a similar manner and may cause difficulty in raising or lowering the undercarriage or in using the brakes.

## Icing in Clear Air

38. HOAR FROST - This term is used to indicate a white, feathery, crystalline formation that can cover the entire surface of the aircraft. It is similar to the ice that occasionally forms on metal surfaces such as car roofs during clear, cool winter nights.

39. Hoar frost forms by sublimation or, in other words, by water vapour which changes directly into ice crystals without going through the water stage. Sublimation occurs when moist air comes into contact with an object at temperatures sufficiently below freezing for ice crystals to form. The lower the humidity, the colder the temperature must be. Therefore hoar frost forms on aircraft when their surfaces are at temperatures sufficiently below freezing and the surrounding air is warmer and moist.

40. Aircraft parked outside on clear, cold winter nights are susceptible to hoar frost. This is because the upper surfaces of aircraft cool by radiation to a temperature below that of the surrounding air. Many aircraft have crashed while attempting to take off due to frost on the aircraft wings. This is because even a light coating markedly increases the stall speed.

41. Hoar frost also forms on aircraft during flight. This can occur when an aircraft that has been flying at below-freezing temperatures descends suddenly into warmer, moist air. This condition lasts until the aircraft warms to the new temperature. Frequently, the frost forms sooner and remains longer, in the vicinity of integral fuel tanks because the fuel warms more slowly than the aircraft structure. Another condition for the development of hoar frost occurs when an aircraft climbs rapidly within an inversion. It is not common in this case, however, for although the aircraft is colder than the surrounding air, the temperature difference is normally not sufficient to cause sublimation.

# Aerodynamic Factors

42. As well as the meteorological factors just described, there are various aerodynamic factors that influence icing.

## Collection Efficiency

43. To a large degree the icing rate depends upon the collection efficiency of the aircraft component involved. "Collection Efficiency" is the fraction of the liquid water collected by the aircraft and it varies

directly with droplet size and aircraft speed, and inversely with the geometric size of the collecting surface. The size of an aircraft component is described in terms of the curvature radius of its leading edge. Those components which have large curvature radii (canopies, thick wings, etc.) collect but a small percentage of the cloud droplets, especially the smaller droplets, and have a low collection efficiency. Components which have a small curvature radii (antenna masts, thin wings, etc.) deform the airflow less, and permit a high proportion of droplets of all sizes to be caught. They have a high collection efficiency. Once ice begins to form, the shape of the collecting surface is modified, with the curvature radius nearly always becoming smaller and the collection efficiency increasing. In general, fighter-type aircraft, because of their greater speed and thinner wings, have higher collection efficiencies than do cargo aircraft (Figure 9-8(a)).

44. The faster the speed of the aircraft the less chance the droplets have to be carried around the airfoil in the air stream so the greater the collection efficiency (Figure 9-8(b)).

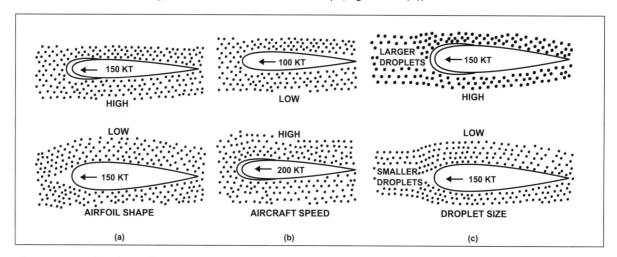

**Figure 9-8 Collection efficiency**

45. Droplet size has an effect on where ice will form. With small droplets, ice formation is limited to the leading edge radius. With medium size droplets ice formation will extend aft of the leading edge radius but not aft of surfaces normally protected by deicing/anti-icing. Ice formation from large droplets will extend aft of the protected surfaces. Ice formation from freezing rain or freezing drizzle can extend aft to the point of maximum component projection into the air stream (Figure 9-8(c)).

**Aerodynamic Heating**

46. "Aerodynamic Heating" is the rise in temperature in the aircraft skin resulting from compression and friction as the aircraft penetrates the air. Compression and friction combine to give the greatest heating at the leading edge of the wing or tail surfaces and decreases to the least heating for the portion to the rear of the mid chord. Ice will not form if the skin temperature is above 0°C. For a given airspeed, ice protection from aerodynamic heating decreases with altitude due to the decrease in air density. In some cases, heating may be sufficient to prevent ice accumulation on leading edges but insufficient to prevent icing elsewhere if the intercepted cloud droplets or precipitation should flow back past the leading edges.

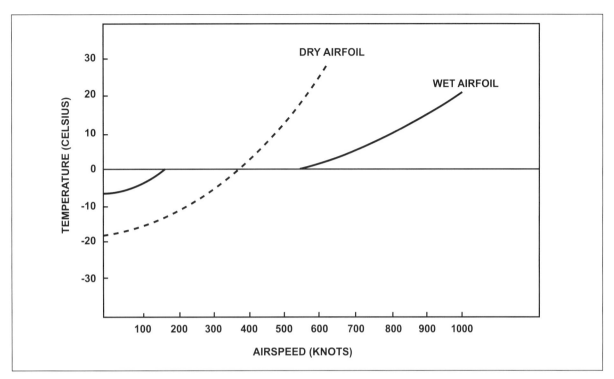

**Figure 9-9 Aerodynamic heating for a wet and a dry airfoil**

47. The amount of aerodynamic heating of a wet airfoil is very much less than a dry one and depends on altitude and the liquid water or ice crystal content of the cloud. Figure 9-9 illustrates this difference for a specific liquid water content at an altitude of 20,000 feet. The temperature of the dry airfoil increases steadily from the outside air temperature as the airspeed increases. The wet airfoil behaves quite differently. The release of latent heat of sublimation as the droplets freeze on the airfoil warms it initially above the outside air temperature. As speed increases aerodynamic heating brings the temperature up to 0°C. At this temperature, heat is required to change the ice to water and then to evaporate the water. This holds the surface at 0°C for a very large range of airspeeds, roughly 400 knots in the case illustrated. With a further increase of speed, the temperature climbs above 0°C. For almost the entire speed range where the temperature is held at 0°C clear ice forms. When speeds are just great enough so that there will be no icing in the leading edge area, there can be icing in the runback area of the airfoil to the rear of the leading edge.

48. Figures 9-10 and 9-11 illustrate the critical temperature for leading edge occurrence icing as a function of altitude and airspeed for an average liquid water content of cloud for subsonic and transonic speeds respectively. Runback ice may form for speeds a little faster than those indicated. For example, no ice will form on leading edges at 12,000 feet at speeds in excess of 300 knots for outside air temperatures warmer than -5°C.

49. In considering overall icing conditions, an airspeed of 500 to 600 knots is required to ensure that no ice will collect.

50. Once ice has formed, aerodynamic heating is very ineffective in removing it. Even few fighter aircraft possess the speed capability required to deice surfaces by aerodynamic heating in the clear dry air above the region where ice has accumulated.

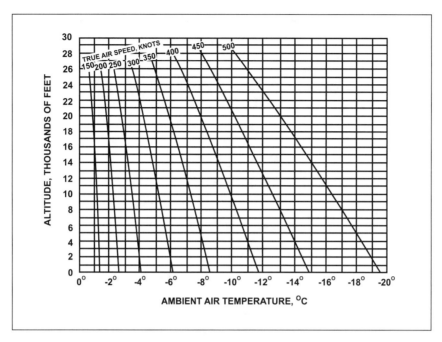

**Figure 9-10 Critical temperature for occurrence of aircraft icing on leading edges – subsonic**

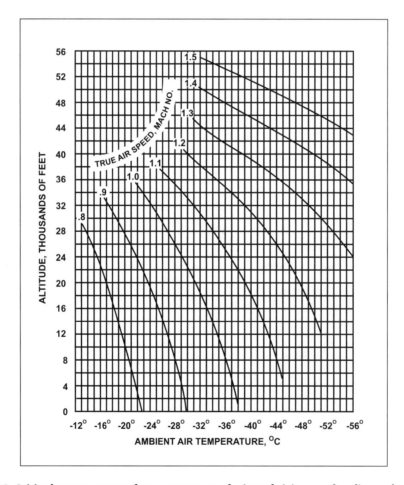

**Figure 9-11 Critical temperature for occurrence of aircraft icing on leading edges – transonic**

# Engine Icing

## Piston Engines - Carburetor Icing

51. Carburetor icing frequently causes engine failure without warning. It may form under conditions in which structural ice could not possibly form. Carburetor icing potential varies greatly among different aircraft and occurs under a wide range of meteorological conditions. If the relative humidity of the outside air being drawn into the carburetor is high, ice can form inside the carburetor in cloudless skies and with the temperature as high as 25°C to 30°C. It sometimes forms with outside air temperatures as low as -10°C. Carburetor ice forms during vaporization of fuel, combined with the expansion of air as it passes through the carburetor. Of the two cooling processes, fuel vaporization causes the greater temperature drop. This may amount to as much as 40°C.

52. This temperature drop can occur in less than a second. Ice will form in the carburetor passages (Figure 9-12) if cooling is sufficient to bring the temperature inside the carburetor down to 0°C or colder and if moisture is available. Ice may form at the discharge nozzle, in the venturi, on or around the butterfly valve, or in the curved passages from the carburetor to the engine.

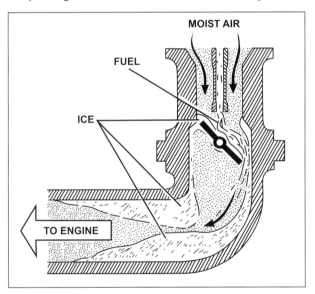

**Figure 9-12 Carburetor icing**

## Powerplant Icing in Jet Aircraft

53. FUEL SYSTEM - Jet fuel has a strong affinity for water. Occasionally, enough water is present to create icing of the fuel system when flying in cold air where the fuel temperature is at or below the freezing temperature of water. This problem usually can be eliminated by the application of heat upstream of the fuel filter or by the addition of a de-icer in the fuel. The heat can be provided electrically, by the use of engine bleed air or by hot engine oil.

54. INDUCTION SYSTEM - Ice forms in the induction system any time atmospheric conditions are favourable for formation of structural icing (visible liquid moisture and freezing temperatures). In addition, induction icing can form in clear air when the relative humidity is high and the free-air temperatures are 10°C or colder.

55. In flights through clouds which contain super-cooled water droplets, air intake duct icing is similar to wing icing. However, the ducts may ice when skies are clear and temperatures are above freezing. While taxiing, during take-off and climb or on approach, reduced pressures exist in the intake system (Figure 9-13). This lowers air temperatures to the point at which condensation and/or sublimation take place, resulting in ice formation. The temperature change varies considerably with different types of engines. Therefore, if the free-air temperature is 10°C or less (especially near the freezing point) and the relative humidity is high, especially if there is fog, the possibility of induction icing definitely exists. When icing of an orifice takes place, ice builds up around the opening, decreasing the radius of the orifice and limiting the air intake. Ice accumulation can become serious within 2 minutes under these critical atmospheric conditions. In most jet aircraft, an airspeed of approximately 250 knots or greater is necessary to help minimize the situation. At airspeeds of 250 knots and above, air is rammed into the intake system rather than sucked into the engine.

56. INLET GUIDE VANES - Icing occurs when the super-cooled water droplets in the atmosphere impinge on the guide vanes and freeze. As a result, blockage of air to the turbine compressor increases with ice build-up. This reduction of air flow to the engine results in a decrease in engine thrust and eventual engine failure. This condition can be alleviated by heating of the inlet components.

57. Damage does not occur from icing in centrifugal flow type turbo-jet engines. However, damage because of ice may occur in axial flow type turbo-jet engines. The shedding of ice accumulations from components ahead of the compressor inlet may cause damage to the engine structure. Small pieces of ice will pass harmlessly through the engine but a large piece of ice could cause severe damage to the engine.

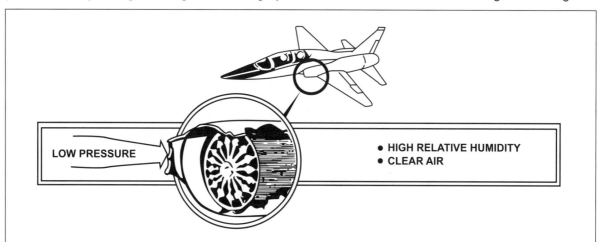

LOW PRESSURE

- HIGH RELATIVE HUMIDITY
- CLEAR AIR

**Figure 9-13 Jet Intake Icing**

58. This shedding can occur from improper use of anti-icing/deicing or from an aircraft letting down into above-freezing temperatures. As the aircraft skin warms, ice can be shed from the rim of the intake, or for aircraft with engines mounted at the rear of the fuselage, from the wings.

59. There are many occasions when the icing situation is quite straightforward and can be handled without difficulty. There are a few occasions, however, when additional factors are present that make conditions very hazardous. The preceding paragraphs have provided you with the necessary information required to recognize when these factors are present. You should be alert and ready for them.

# Summary - Chapter 9

- Water droplets can exist as super-cooled water at temperatures below 0°C.

- The intensity of icing is greatest in an area of high liquid water content.

- High liquid water content in cloud occurs when temperatures are near freezing and when droplets are large.

- The liquid water content of cloud can decrease rapidly with the appearance of ice crystals.

- Ice occurs as rime, clear and mixed. Clear ice presents the greatest hazard.

- Clear ice forms from the slow freezing of large droplets, rime from rapid freezing of small droplets.

- The intensity of ice is described as: trace, light, moderate and severe.

- Convective cloud tends to produce clear ice. The icing will be particularly serious in developing cloud and can occur down to -25°C. It is of limited horizontal extent.

- Icing in layer cloud is normally less serious but of greater horizontal extent. It can be serious with:

  - Stratocumulus over water, particularly near the cloud top.

  - Any layer cloud formed due to rapid forced ascent (intensifying low or front, strong orographic lift, lee wave cloud).

- Freezing rain is most often associated with a warm front and can cause severe clear ice.

- Freezing drizzle occurs under stratus and can cause severe clear ice.

- Frost and wet snow also pose icing hazards.

- Aerodynamically, the greater the collection efficiency the greater the icing hazard.

- Collection efficiency is large for sharp leading edges, high speeds and large water droplets.

- Aerodynamic heating can keep the aircraft skin above freezing and prevent ice, but speeds in excess of 500 knots may be required.

- Ice also forms in piston and jet induction systems and in fuel.

**9** CHAPTER

# Chapter

# 10

Your visibility from the cockpit will frequently be the most important factor with which you must contend.

# CHAPTER 10

## VISIBILITY

1. The visibility required for various types of flying varies enormously. When a pilot is first learning to fly, the horizon, which may be 100 miles away, should be visible. VFR (Visual Flight Rules) requires about three miles visibility so that visual separation can be maintained from other aircraft. IFR (Instrument Flight Rules) landing requires that enough of the runway be seen from about half a mile away so that a visual landing can be made.

2. How well you can see depends on several factors such as the presence of fog, haze, cloud or precipitation, whether you are looking into the sun or away from it and even whether it is daylight, twilight or night. These and other factors will be described in this chapter.

3. Ground level, air-to-ground and air-to-air visibilities are all important when flying. Ground level visibility is routinely provided to you in aviation weather reports and forecasts by the Weather Service. Air-to-ground and air-to-air visibility may occasionally be provided by reports from other aircraft. However, you will frequently have to deduce them from reports or forecasts of ground level visibility. Most cases of reduced visibility are associated with inversions since they act as lids that confine particles or water droplets suspended in the air to the layers below the inversions.

### Ground-Level Visibility

4. A weather observer uses a prominent object viewed against the horizon for estimating daytime visibility. At night-time an unfocussed light of moderate intensity is used in an attempt to provide visibility equivalent to the daytime visibility. There is a time at twilight when visual difficulty is encountered because ground features are fading out, yet it is not dark enough for lights to show up brightly. You will encounter this difficulty when you are flying.

### Prevailing Visiblity

5. The prevailing visibility is provided for aviation by the Weather Service. It is the maximum visibility common to sectors comprising one half or more of the horizon circle as viewed from the observing site at eye level. It is provided in statute miles.

6. The interpretation of the prevailing visibility is illustrated in Figure 10-1. The maximum visibility common to half or more of the horizon circle is 1 mile. There can be complications in providing visibility in this manner which you should be aware of. Note that in this example, although the prevailing visibility is 1 mile, the visibility on the approach to runway 25 is 3 miles but to runway 07, it is only 1/4 mile. Important variations such as this may be provided to you when you are given the prevailing visibility.

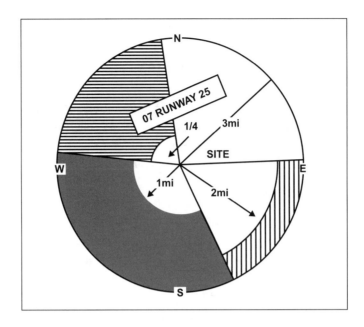

**Figure 10-1 Prevailing visibility**

## Runway Visual Range (RVR)

7.  For landing and take-off under instrument flight conditions, the prevailing visibility is not of as much importance as the visibility within the runway environment itself. Within this environment, the requirement is that the runway lights, rather than the ground features, be visible. "Runway Visual Range" (RVR) is an assessment of the maximum distance in the take-off or landing direction that the runway lights can be seen. It is measured by an instrument called a "forward scatter meter" located near the runway threshold. It uses the forward scatter principle to measure the atmospheric extinction coefficient and extrapolates this into a visibility.

8.  To calculate RVR, three factors must be known. The first is the visibility provided by the forward scatter meter, the second is the ambient light level and the third is the runway light intensity (controlled on request by the ATC controller). Those three factors are input into equations that are processed by computer. The result is the RVR provided in feet.

9.  On some occasions the RVR will not accurately represent what the pilot sees. Those conditions generally occur during the day with shallow fog or with a snow surface. The glare caused by the sun reflecting on those surfaces will affect the pilot's view. In those situations, the use of prevailing visibility would be more appropriate (more details in Chapter 17).

## Air-to-Ground Visibility

10. Prevailing visibility and Runway Visual Range are horizontal visibilities near the earth's surface and they may be quite different than your visibility from the cockpit when looking down at the ground during flight. The surface visibility can be quite seriously reduced in fog or blowing snow and yet from an aircraft at altitude there may appear to be only a very slight reduction in visibility. An example is shown in Figure 10-2. From a position well above the fog, the aerodrome can be seen relatively clearly through a small thickness of fog. On descending to the level of the fog, however, a much greater thickness must be seen through and the aerodrome may disappear from sight. On the other hand, the diffuse reflection from a low sun on the top of a haze layer may seriously reduce the air-to-ground visibility to values less than the ground visibility as illustrated in Figure 10-3. When you are given the ground visibility, you must learn to anticipate what your air-to-ground visibility is likely to be. It will depend on the weather factors present.

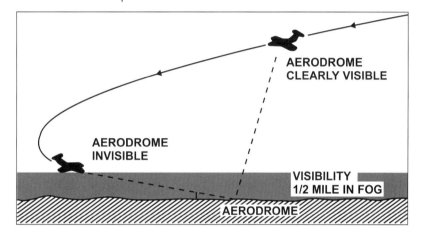

**Figure 10-2 The effect of shallow fog on visibility**

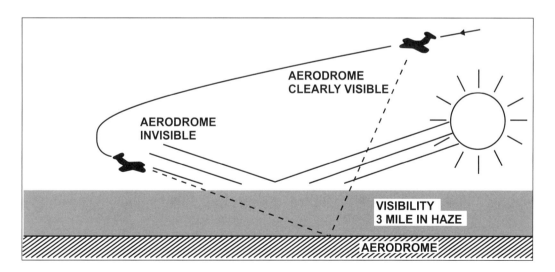

**Figure 10-3 Visibility effects of sunshine on a haze layer**

## Slant VIsual range (SVR)

11. Pilots making an approach for landing, particularly during instrument approaches, are concerned with the distance that they can see the runway landing aids in front of them. This visibility is called the "Slant Visual Range". It is probably the most vital weather information required for landing. Unfortunately, there is great difficulty in either estimating the slant visual range or measuring it from the ground, so it is not provided. The runway visual range provides the best clue as to what the slant visual range will be. However, other weather information such as precipitation and the prevailing visibility will help.

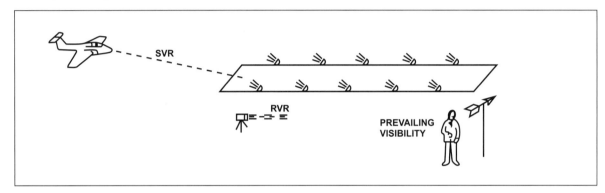

**Figure 10-4 Prevailing visibility, SVR and RVR**

## Air-to-Air Visibility

12. Your air-to-air visibility depends mainly on the presence of cloud or precipitation. The factors that are described in the following paragraphs can also reduce air-to-air visibility.

# Causes of Reduced Visibility

## Lithometers

13. Lithometers are dry particles suspended in the air and include haze, smoke, sand and dust.

## Haze

14. Haze consists of tiny dry particles suspended in the air that produce a bluish colour when viewed against a dark background. Although haze may occur at any level in the troposphere, it is more common in the lower few thousand feet. The top of a haze layer which is confined by a low-level inversion has the appearance of a horizon when viewed from above. In this type of situation, the haze may completely obscure the ground in all directions away from the vertical.

15. Dense haze may reduce the visibility to less than three miles, with the air-to-ground visibility generally being less than the surface visibility. The visibility in haze is lower when looking towards the sun than away from it but is better when looking towards the moon than away from it.

## Smoke

16. Smoke consists of fine ash particles suspended in the air. When smoke is present, the disc of the sun at sunrise and sunset appears very red. Smoke reduces visibility in a manner similar to haze, although smoke from forest fires frequently is concentrated in layers aloft with good visibility beneath. Near industrial areas with a low-level inversion, the smoke concentrates in the lower levels, greatly reducing the visibility. You should recall that smoke particles are nuclei for condensation of water vapour so smoke and fog or stratus are often mixed.

## Sand and Dust

17. When the air is unstable, the wind strong and the soil loose and dry, dust and sand can be blown up into the air. Dust is finer than sand and may at times be lifted to 10,000 or 15,000 feet, whereas the sand will be held down to 50 or 100 feet. In severe sand or dust storms, visibility can drop to near zero.

## Visibility in the Stratosphere

18. There are special problems in the stratosphere. In the absence of any objects on which to focus, the eye adjusts itself to a focus midway between distant and near vision. Because of this, an aircraft that should be plainly visible can escape detection. The air is exceptionally clear in the stratosphere with no haze or cloud and this leads to brilliant dazzle from the sun. This, too, can create visibility problems.

**10** CHAPTER

# Precipitation

## Rain and Drizzle

19. Liquid precipitation can, on its own, reduce visibility, but it also causes a further reduction of visibility from the cockpit as it streams across the windscreen. Without the windscreen effect, rain will seldom reduce visibility below one mile except in brief heavy showers. Drizzle can restrict visibility to a much greater degree. Drizzle falls in stable air and, therefore, is frequently associated with fog or smoke.

20. Rain or drizzle streaming across a windscreen reduces the visibility from the cockpit and also causes a "Refractive Error." This error is such that objects appear lower than they actually are. For instance, a hilltop at half a nautical mile ahead of an aircraft could appear to be about 260 feet lower than it really is.

## Snow

21. Snow affects visibility much more than rain or drizzle and can easily reduce it to less than one mile. There are occasions during flight when it is difficult to see that snow is falling from cloud ahead. You can unexpectedly enter the snow and get into serious difficulty because you lose all visual references.

### Blowing Snow

22. If snow is fine and dry it can easily be lifted by the wind. It is usually confined to the lower 100 to 300 feet; however, this depends on the strength of the wind and the stability of the air. During or after a fresh snowfall and with brisk winds, the surface visibility can be reduced to less than half a mile. Under the continued influence of strong winds, the snow becomes compacted and the blowing condition decreases. If an ice crust forms on the snow due to mild temperatures, it will not lift to blowing snow even with strong winds.

# Fog

23. Fog is one of the most common and persistent weather hazards encountered in aviation so it is of particular importance to you. It is cloud, based on the ground, and is composed of either water droplets or super-cooled water droplets, ice crystals or a mixture of ice crystals and supercooled water droplets. Ideal atmospheric conditions for the formation of fog are a high relative humidity, an abundance of condensation nuclei and some process that will cause condensation to occur in the lower levels of the atmosphere. Fog is most prevalent in coastal areas where moisture is abundant and in industrial areas where it can form at less that 100% relative humidity because of the abundance of condensation nuclei. The names given to various types of fog are based on the way that they are formed.

### Radiation Fog

24. In Chapter 3, "Cooling of the Atmosphere," radiation cooling was described. Due to the radiation of heat from the earth's surface at night, the surface layer of air cools and an inversion forms. The amount of cooling depends on several factors. A land surface cools markedly, whereas a sea surface cools very little. The temperature drop is larger with clear skies than with cloud cover. The depth of the layer cooled is increased by mechanical turbulence if there is a wind. However, if the wind is very strong, the surface temperature drop is small because the cooling effect is spread through a deeper layer of air. Air cooled through radiation tends to drain into low-lying areas.

25. When the air cools to its dew point, condensation will occur. On occasion, this will only cause dew or hoar frost to form; however, with a slight mixing of the air, radiation fog will form.

26. The ideal conditions for radiation fog are: a land surface, light winds, clear skies and moist air. An abundance of condensation nuclei is an important additional factor. These conditions are found in a high pressure area composed of a maritime air mass and near a source of industrial pollution. The fog will form first in valleys and low-lying areas as the cooler air drains into them. For an hour or so after sunrise, the earth's surface continues to cool until incoming radiation exceeds outgoing radiation. At the same time, just after sunrise, the solar radiation begins to stir the air slightly. This slight mixing may cause fog to form if it has not formed previously, or if it has already formed, to make it thicker and to further reduce the visibility.

CHAPTER **10**

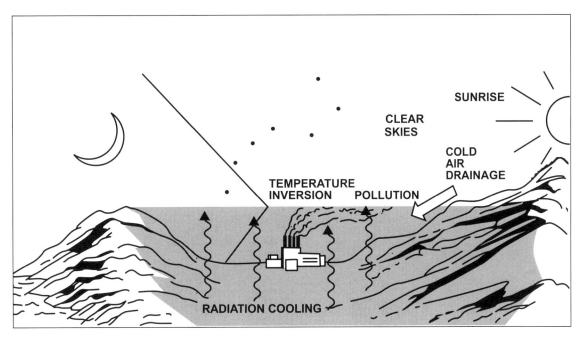

**Figure 10-5 Radiation fog**

27. During the morning, when more heat is received from the sun, the earth's surface and then the air in contact with it begins to heat. The fog will then start to dissipate from the earth's surface upward, becoming very patchy, then forming stratus-fractus cloud and finally dissipating. The fog does not burn off from the top down, but rather, from the bottom up. If cloud should move over an area of fog near sunrise, this could delay or prevent the dissipation of the fog.

28. Radiation fog is normally only a few hundred feet thick. It has a sharp top with clear skies above. It tends to be patchy especially when it is forming or dissipating and can reduce the visibility to zero. It generally takes a thickness of about 300 feet of fog to obscure the sun from the earth's surface.

29. The question of possible diversion becomes important in fog conditions. In a radiation fog situation, areas free of fog could be found on hilltops or hillsides, in coastal areas with an onshore wind or in windy, cloudy areas.

**Advection Fog**

30. Advective cooling is another method described in Chapter 3 by which the lower layers of the atmosphere are cooled. "Advection Fog" forms due to the cooling that results from air moving over a surface colder than itself. Advection fog deepens as the wind speed increases up to about 15 knots. With winds stronger than this, the fog will lift to stratus except where the amount of cooling is extreme. For instance, fog will persist with very strong winds in air moving northward off the Gulf Stream onto the Labrador Current near the East Coast.

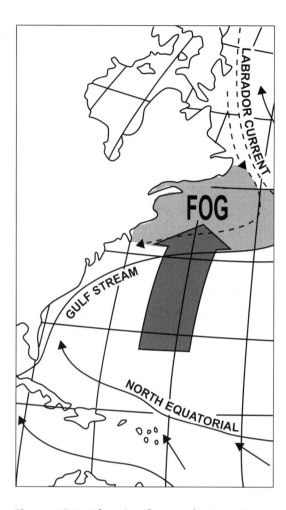

**Figure 10-6 Advection fog on the East Coast**

31. Advection fog forms as a result of an air flow of air from: warm land to cold sea, from warm sea to cold sea and from air cooled as it moves from south to north. Over oceans or large lakes, there will be no improvement from night to day. The fog will persist unless there is a change in the air mass or the wind direction. Over land, if the sun is not blocked by cloud, there may be sufficient heating to cause the fog to dissipate during daylight.

32. Alternate aerodromes for an advective fog situation can frequently be found in the lee of hills or mountains where subsidence heating will dissipate the fog (Figure 10-7).

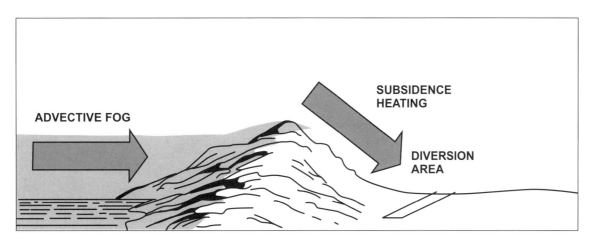

**Figure 10-7 Diversion area for advection fog**

## Upslope Fog

33. Upslope fog is simply cloud resting on the ground that has formed due to the cooling of air by expansion as it moves up a slope. With winds of moderate strength the fog can persist. However, with stronger winds, mechanical turbulence will lift it to stratus or stratocumulus. Upslope fog will persist until there is a change to a drier air mass or the wind direction changes. If the sun can penetrate the fog, there may be a slight improvement during the day.

34. Alternate aerodromes for upslope fog conditions can be found in the lee of hills or mountains where subsidence heating is occurring or where the air flow is no longer riding up the slope.

## Steam Fog or Arctic Sea Smoke

35. In the previous examples, fog formed due to the cooling of air, however "Steam Fog" forms when water vapour is added to the air. In order for this to happen, there must be a sharp low-level inversion and the air temperature must be low so that the air can be easily saturated. Ideal conditions exist in the Arctic when extremely cold air flows over open ocean leads. The air becomes unstable in a very shallow layer and the evaporation from the much warmer water surface quickly saturates the air. As a result, fog forms which may rise vigorously and drift hundreds of miles downwind. Coastal airports can be seriously affected (Figure 10-8).

**Figure 10-8 Arctic sea smoke**

## Ice Fog (Combustion Fog)

36. Ice fog is a man-made phenomenon that occurs with very cold winter temperatures. It is formed of ice crystals and is caused by the addition of water vapour to the air through fuel combustion. With very low temperatures, a strong surface inversion is common and the air beneath the inversion becomes easily saturated. The density of ice fog will build up over a period of time if the winds are light so that the visibility can be reduced to near zero. Combustion from furnaces, aircraft engines and automobiles are all sources of moisture for ice fog.

**Figure 10-9 Aircraft causing ice fog**

## Frontal Fog (Precipitation Fog)

37. With light winds, fog may form instead of stratus (The formation of stratus in precipitation was described in Chapter 8.) in the continuous precipitation preceding a warm front or trowal. It will advance with the system and clearing will occur as the frontal weather moves off.

**Figure 10-10 Frontal fog**

## White-Out

38. "White-out" is an atmospheric optical phenomenon in which the observer appears to be engulfed in a uniformly white glow. Neither shadows, horizon, nor clouds are discernible. Orientation and depth perception are lost. Only very dark, nearby objects can be seen. White-out occurs over an unbroken snow cover beneath a uniformly overcast sky, when the light from the sky is about equal to that emanating from the snow surface. The actual visibility may be excellent.

39. A hazard of white-out is the pilot not suspecting the phenomenon because he/she is in "Clear Air." In numerous white-out accidents, pilots have flown into snow-covered surfaces unaware that they have been descending and were confident that they could "see" the ground. Consequently, whenever a pilot encounters the white-out conditions, or even suspects them, he/she should immediately climb if at a low level, or level off and turn towards an area where sharp terrain features exist. The flight should not proceed unless the pilot is prepared and competent to traverse the whiteout area on instruments.

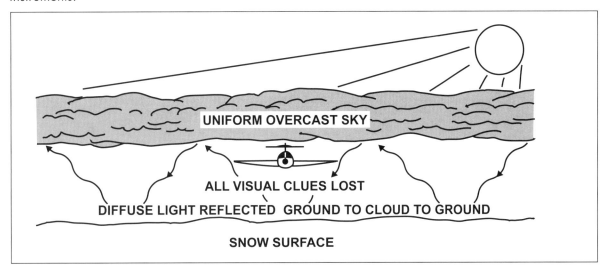

**Figure 10-11 White-out**

## Visual Horizon

40. It is sometimes necessary to know how high you must fly in order to see distant objects on the ground. For flat terrain, the distance of the horizon (s) in nautical miles from a height (h) in feet is given by the expression:

$$S = 1.14\sqrt{h}$$

# Summary - Chapter 10

- Prevailing visibility is the ground visibility provided by the Weather Service in reports and forecasts.

- RVR is a measured visibility indicating the maximum distance that runway lights can be seen.

- Air-to-ground visibility may vary markedly from the surface visibility because of weather factors.

- SVR is the distance landing aids can be seen on the approach. RVR provides the best assessment of what the SVR will be.

- Air-to-air visibility can be restricted because of cloud, haze and precipitation. Dazzle from the sun and difficulty in focussing on distant objects cause difficulty in the stratosphere.

- Haze and smoke can reduce visibility to less than three miles. Visibility is lower when looking into the sun or away from the moon. Smoke can occur as a layer aloft or as a surface-based layer.

- Blowing sand, snow and dust can restrict visibility to near zero. Blowing snow and blowing sand are generally restricted to the lower levels of the atmosphere.

- Snow affects visibility more than drizzle, which in turn affects it more than rain. Rain and drizzle streaming across the windscreen cause a further reduction in visibility and a refractive error.

- Areas of snowfall may be difficult to see while flying and may be flown into unexpectedly.

- Radiation fog requires a land surface, light winds, clear skies and moist air. It forms at night and dissipates during the morning.

- Alternate aerodrome areas for radiation fog include high ground, coastal areas with an onshore flow and cloudy, windy areas.

- Advection fog occurs when moist air moves over a colder surface. It persists day and night.

- Alternate aerodrome areas for advection fog can be found in the lee of hills or mountains.

- Upslope fog forms from moist air moving up a slope. It persists day or night.

- Alternate aerodrome areas for upslope fog can be found in the lee of hills or mountains, or where air has turned to a downslope flow.

- Steam fog occurs when very cold air flows over much warmer water.

- Ice fog is caused by fuel combustion under very cold conditions.

- Frontal fog occurs in the precipitation area of fronts, particularly warm fronts.

- White-out occurs under an overcast sky and over a snow surface with bright sun shining on the cloud top.

- The formula $S = 1.14\sqrt{h}$ gives the distance to the horizon in nautical miles.

# Chapter

# 11

Low-level flying is complicated by turbulence and abrupt changes in wind.

# CHAPTER 11

## BOUNDARY LAYER WINDS AND TURBULENCE

1. The winds and turbulence that will be described in this chapter are those in the lower portion of the atmosphere that are affected by the earth's surface and terrain. The winds within this layer are extremely variable and may bear little relationship to the isobaric pattern. Just above this layer the winds become geostrophic, flowing parallel to the surface isobars at a speed proportional to the isobar spacing.

### Wind Shear

2. When operating an aircraft, you must adapt your flying to the winds that are present. For example, take-offs and landings are generally into wind, but if a crosswind exists, special techniques are required. Of particular importance within the boundary layer are abrupt changes that occur in the wind field and these must be compensated for when flying. These changes can be in both speed and direction and are called "Wind Shears."

3. Wind shear can be described as the difference in wind between two separate points. The points may be spaced vertically if vertical wind shear is being described, or horizontally if horizontal wind shear is being described. Our concern in aviation is the difference in the wind along the flight path of an aircraft, whether the aircraft is flying straight and level, climbing or descending.

4. If an aircraft were heading into a 30-knot head wind and this gradually changed because of a change in the wind pattern to a 30-knot tail wind, nothing significant other than a change in the ground speed would occur. If the change happened abruptly, however, there would be a rapid change in the airspeed. Because of the large mass of an aircraft, it has considerable inertia. If the wind changes at a rate greater than the aircraft can accelerate or decelerate because of its inertia, the airspeed will change abruptly. For large and heavy aircraft it will take an appreciable time, possibly a minute, before the airspeed will return to its original value. Even for light aircraft the delay in regaining airspeed can be significant. Ground speed also changes, but much more slowly than the airspeed.

### Classification of Shear

5. Shears can be classified according to the effect on the aircraft. An increased performance shear (also called a headwind shear) occurs when the shear causes the airspeed to increase. A decreased performance shear (tailwind shear) occurs when the shear causes the airspeed to decrease. An increased performance shear results from a rapidly increasing head wind or a decreasing tail wind. A decreased performance shear results from a rapidly decreasing head wind or an increasing tail wind. The effect of shear on an aircraft is particularly important during take-off and landing because it can cause stalls, undershoots or overshoots depending upon the situation.

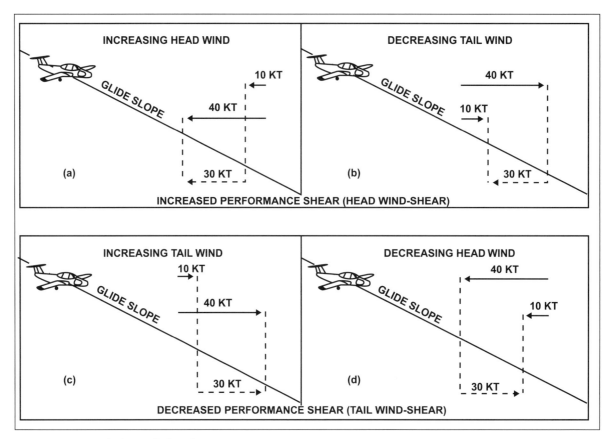

**Figure 11-1 Wind shear defined**

6.  In Figure 11-1(a), you are in an aircraft on the glide slope and pass quickly from a 10-knot head wind into a 40-knot head wind. This is a 30-knot increased performance shear and the airspeed will increase. In Figure 11-1(b) you pass from a 40-knot tail wind to a 10-knot tail wind. This, too, is a 30-knot increased performance shear and the airspeed will increase. In Figure 11-1(c), you fly from a 10-knot tail wind into a 40-knot tail wind. This is a decreased performance shear and the airspeed will decrease. In Figure 11-1(d), you fly from a 40-knot head wind into a 10-knot head wind. This again is a decreased performance shear and the airspeed will decrease. In each case after the airspeed has changed abruptly, it will return to its original value, but slowly.

7.  There can also be crosswind shears and up and down drafts that affect the aircraft. An abrupt crosswind shear will make an aircraft weathercock into the new wind. An abrupt downdraft causes a brief decrease in the wing's attack angle with a resulting decrease in lift. An updraft causes an increase in the attack angle and an increase in lift for a short period. With an updraft, however, the attack angle may be increased beyond the stall angle causing a marked decrease in the lift being produced.

8.  There is frequently noticeable turbulence in a shear zone, but this is not always the case (Figure 11-2). The absence of turbulence cannot be counted upon as a sure indication that shear is not present. Turbulence in a shear zone, in fact, will tend to dissipate the shear since it causes the air in one flow to mix with that in the other combining the two velocities into one. Unless the force creating the shear is greater than the turbulent forces trying to dissipate it, turbulence will gradually reduce the amount of shear.

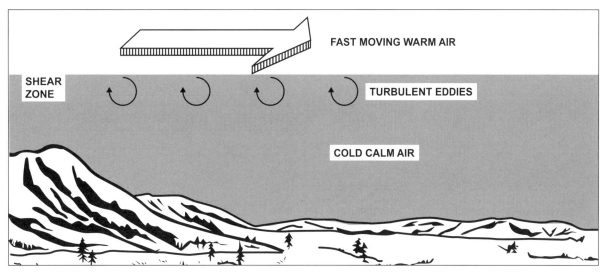

**Figure 11-2 Shear turbulence**

## The Effect of Shear

9.  Figure 11-3 illustrates the effect of increased performance shear and decreased performance shear during an approach to land. In Figure 11-3(a) you abruptly fly from a 10-knot head wind into a 30-knot head wind. This is a 20-knot increased performance shear and the airspeed will increase. Since you have entered an area of stronger head winds, it would be logical to assume that you would undershoot unless you add power. The resulting increase in airspeed, however, has a very pronounced effect on the path that the aircraft follows.

10. The lift produced by a wing is given by the equation $L = 1/2dV^2C\ S$, where d is the density of the air, V is the true airspeed, C is the coefficient of lift and S is the surface area of the wing. The lift varies as the square of the true airspeed so the airspeed has a very marked influence on the amount of lift produced.

11. In Figure 11-3(a), you encounter the shear at a low altitude just prior to touchdown. Because of the increase in airspeed as the aircraft passes through the shear, lift increases markedly and the aircraft rises above the glide path. The aircraft is now high and fast and a long float and possible overshoot will result.

12. In Figure 11-3(b) you encounter the increased performance shear at a higher altitude during the approach. After rising above the glide slope you recapture it and you must then adjust your throttle to more power than was originally used because you are now in an increased head wind.

13. In Figure 11-3(c) you have encountered a decreased performance shear just prior to touchdown Because of the decreased airspeed, lift decreases and you sink below the glide path. You are now low and slow and may undershoot or stall onto the ground.

14. In Figure 11-3(d) you encounter the decreased performance shear at a higher altitude during the approach. After sinking below the glide path, you recapture it and you must then readjust the power to something less than the original power because of the decreased head wind. From these examples,

you can see that the actions that are required depend on the height that the shear is encountered in relation to the runway, and if it is an increased performance or decreased performance shear. Shear encounters on take-off can also occur and may be just as hazardous as on landing. Aircraft have struck the ground and crashed in a decreased performance shear situation.

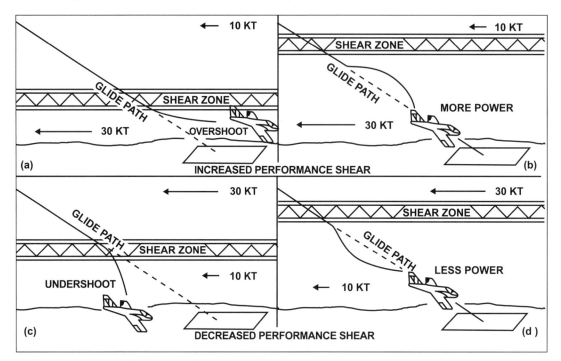

**Figure 11-3 Effect of shear on the approach**

15. You will encounter shears many times during your flying. Generally you will make corrections for them with little difficulty and, in fact, you may not even be aware that you have flown through a shear zone. There are occasions however, when the shear is extreme and very rapid and positive corrective action is required to avoid an accident. In these cases, if you recognize the possibility of shear ahead of time, your reactions will be more prompt and recovery more assured. You should be particularly aware that the effect of shear is quite different from that of gusty winds. Gusts are very transitory so that any increases in head wind or tail wind components are almost immediately followed by decreases.

## Frontal Shear

16. Earlier in this manual, frontal structures were described. Wind shifts associated with frontal surfaces do not often produce shear significant for aircraft taking off or landing. They normally contain gradual wind changes spread over a distance in the order of 50 to 100 miles horizontally and 3,000 or 4,000 feet vertically. While they do contain shears, they are generally of an intensity that can easily be handled.

17. On occasion, however, sharp shears do occur. A measurement made by a research aircraft through a cold front indicated a 30-knot wind shear through a frontal inversion that was only 200 feet thick. A study done by Northwest Airlines has provided evidence that if a cold front has a 5°C or greater difference between the air masses, or if it is moving faster than 30 knots, noticeable shear can be expected. So, while strong shears may be unusual with fronts, they can occur. You should be on the alert for shear if you are doing a letdown or take-off through a frontal surface.

18. You can encounter shear with warm fronts as well as with cold fronts. In Figure 11-4 you are initially on the glide slope in the warm air with the warm air wind flowing from the southwest. As you descend, you cut through the frontal surface into the cold air where the wind is from the southeast. This is an increased performance shear. It is well to remember that in frontal diagrams, the vertical scale is invariably exaggerated. The typical warm front slope of 1 in 150 gives an angle of only about 0.4° between the earth's surface and the sloping frontal surface. A cold front with a typical slope of 1 in 50 will have an angle of about 1.2°. A glide slope of 2 1/2° to 3° will cut through a frontal surface if it should be lying over the approach.

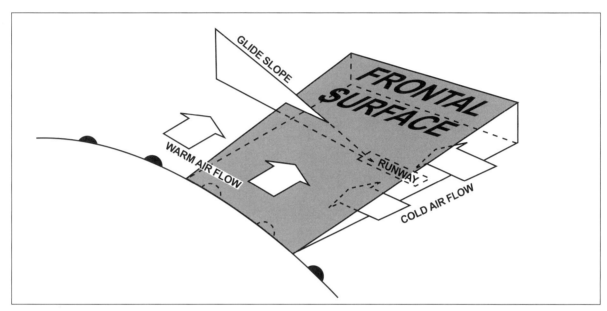

**Figure 11-4 Frontal shear**

19. The paragraphs that follow in this chapter describe various topographical features of the earth's surface and the air flow that results. Mention is made of shear in the cases where it can be significant. Shear also occurs in association with thunderstorms and a phenomenon called "Mountain Waves." These are discussed later in Chapter 16, which is devoted to the effects of low-level wind shear on instrument approaches and take-offs.

## Stability and the Diurnal Variation of Wind

20. Given the same isobaric pattern and pressure gradient, surface winds are usually stronger and gustier during the day than during the night. In unstable air there are ascending and descending currents that provide a vertical link between the winds at different levels in the convective layer. In stable air, particularly in an inversion, this link does not exist so that winds can be very different above than below the inversion.

21. A typical example occurs with the night-time inversion (Figure 11-5). As the sun sets and the nocturnal inversion develops, the link between the air at the lowest levels that is affected by the earth's surface and the free flowing air above disappears. The surface air becomes more strongly influenced by the frictional effects of the earth's surface so that the surface wind decreases in speed and backs in

direction (see Chapter 5, paragraph 20). This process starts at the surface and works upward for several hundred feet as the inversion deepens during the night.

22. As the sun rises, instability develops, the inversion breaks down and vertical currents begin. This causes a general mixing in which the faster air aloft is brought down to the surface and the slower surface air is carried aloft. This causes the surface wind to increase in speed, veer in direction and become gusty. It reaches a maximum speed during the late afternoon at the time of maximum temperature (Figure 11-5). Diurnal variations such as this will not occur over large lakes or oceans since no nocturnal inversion develops over them. For this reason there is little diurnal variation of wind over large bodies of water.

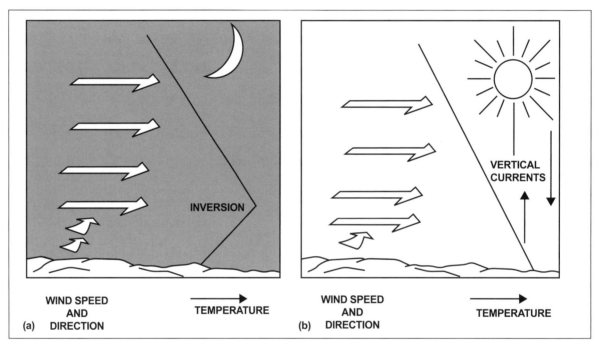

**Figure 11-5 The diurnal variation of surface wind**

## The Flow Over Hills and Mountains

23. With stable air, the flow tends to be laminar, flowing smoothly over terrain with little turbulence but with marked vertical wind shear. The wind speed will increase over the crests as shown in Figure 11-6(a). If the wind speed is fast, or the air flows over a sharp ridge, the laminar flow will break down and turbulent eddies will form on the lee side (Figure 11-6(b)). A large stationary roll eddy may form as shown in Figure 11-6(c) resulting in a moderate to strong upslope wind opposite in direction to that flowing over the rim.

24. Unstable air does not have a laminar flow over hills and mountains but breaks up into chaotic and turbulent convection with strong up and down drafts.

(a) LAMINAR FLOW

(b) TURBULENT FLOW

(c) UPSLOPE FLOW OPPOSED TO THE GENERAL AIR FLOW

**Figure 11-6 Flow of stable air over a ridge**

## Density

25. Differences in air density play a part in the development of local winds. Since density is largely controlled by the air temperature, the heating and cooling of the air near the earth's surface is important. Cold, dense air will flow down slopes and valleys similar to a flow of water. Warm, buoyant air, on the other hand, will rise up slopes and valleys. Radiational heating and cooling, and cooling of the air over ice and glacier surfaces are the major factors producing local variations in the density of the air.

## Topographical Effects

26. Winds and turbulence that develop in hilly and mountainous terrain are of particular importance for light aircraft, helicopter and low-level fighter operations. In mountainous areas where the performance of light aircraft or helicopters is marginal, the location of upslope and downslope winds can be critical. Fighter aircraft operating at high speed in hilly or mountainous areas can easily be overstressed due to turbulence.

## Funnel Winds

27. Figure 11-7 illustrates the effect that a mountain range, or to a lesser extent, a range of hills, or even buildings, can have on air flowing against them. Particularly if the air is cold and stable, it will be largely deflected by the obstruction and will not rise over it. Under these conditions, if an opening such as a gorge or valley exists, the wind will funnel through at greatly increased speed. "Funnel winds" such as these flow out of the mountain valleys and over the ocean on the West Coast reaching speeds of up to 80 knots and creating hazardous shear and turbulent conditions.

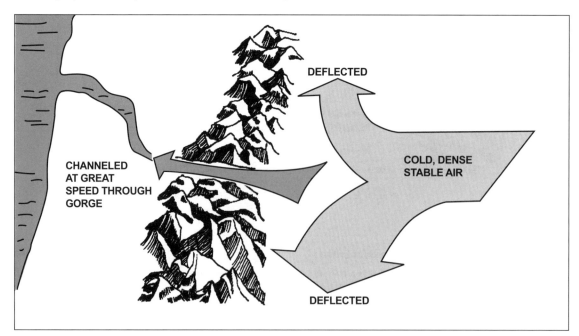

DEFLECTED

COLD, DENSE
STABLE AIR

CHANNELED
AT GREAT
SPEED THROUGH
GORGE

DEFLECTED

**Figure 11-7 Air being deflected at a mountain range and channelled through a gorge**

# Slope and Valley Winds

## Anabatic Winds

28. An "Anabatic Wind" is a wind that flows up a slope or a valley during the day. Air adjacent to a sun-facing slope becomes warmer and less dense than the air at the same level some distance from the slope. The warm air becomes buoyant and floats up the slope. The surface heating develops instability which causes the flow to be turbulent. The depth of the upslope wind increases with height and the air can become very turbulent at the top of the hill or ridge (Figure 11-8). For the same reason, air tends to flow up a valley during the day rather than along isobars.

**Figure 11-8 Anabatic wind**        **Figure 11-9 Winds over sunny and shaded slopes**

29. Figure 11-9 illustrates a valley with one slope facing the sun and the other slope in the shade. In situations such as this, there will be an upslope wind on the sunny side and up the valley and a downslope wind on the shaded side.

## Katabatic Winds

30. A downslope wind is called a "Katabatic Wind." It may consist of either a warm or a cold flow of air down a slope and may develop into an extremely strong wind with dangerous shears.

31. Air cooled by radiational cooling at night becomes dense and is pulled by gravity down the natural pathways of hills and mountains to collect in the valleys and then flows down the valleys out onto the plains or oceans. The flow tends to be shallow and laminar (Figure 11-10)

**Figure 11-10 Katabatic winds**          **Figure 11-11 Glacier winds**

## Severe Downslope Wind

### Glacier Winds

32. Katabatic winds can develop to hazardous proportions if the cooling is extreme. This can occur over glaciers where shallow winds of 80 knots or more can form (Figure 11-11). Since the cooling in this case is caused by the underlying ice, the winds blow downslope both day and night. At times, the flow is pulsating with the cold air building up to a critical point then being released in a rush down the slope.

### Chinooks

33. The chinook of Alberta is an example of a warm katabatic wind. The chinook is most common between September and April. A deep layer of moist air from the Pacific is forced upward across the mountains. As it rises it cools dry adiabatically until it is saturated and then it cools moist adiabatically. The precipitation that falls out of the cloud on the western slopes of the mountains is lost to the air mass, so it becomes drier. Upon descending the leeward slopes the air is heated at the dry adiabatic lapse rate and arrives at lower elevations both warmer and drier than it was originally.

34. In Figure 11-12, air originally at 15°C cools to 12°C by 1,000 feet where it becomes saturated. It continues cooling moist adiabatically to reach -2°C at the mountain top at 10,000 feet. It then descends 10,000 feet, heating at 3°C/1,000 feet to reach the base of the mountain at 28°C. Air originally at 15°C has warmed to 28°C, and has become much drier due to the precipitation lost on the windward slopes.

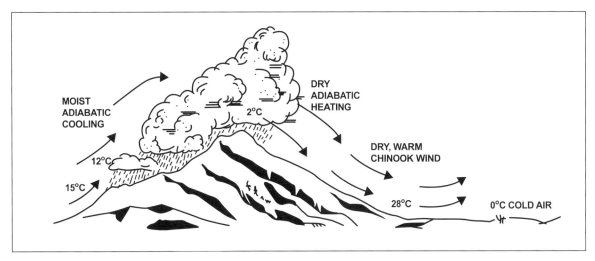

**Figure 11-12 The chinook**

35. At the time of year that the chinook is common, cold Arctic air frequently covers the Prairies and lies as a shallow layer up against the mountains. This air is dense and will normally hold the warm chinook air above it. There is a phenomenon, however, called a "Mountain Wave," that is described in Chapter 16, that can abruptly shift the cold air away from the mountain so that the warm air strikes the surface as a hot, dry blast. Temperatures can jump by 20°C in a few minutes. Turbulence and shears can be extreme in a chinook.

## Land and Sea Breezes

36. As the name suggests, these are breezes and they do not develop into the severe winds that have been described previously. The temperature of land masses rises and falls more rapidly than do water surfaces due to insolation and terrestrial radiation. The result is, the land is warmer than the sea during the day and cooler during the night. This difference in temperature is acquired by the overlying air. It is greatest during the summer and when the winds are light. Because of warmer temperatures over the land during the day, the pressure decreases and becomes less than that over the water. The colder air over the water moves towards the lower pressure, forcing the warm air over the land upwards. A circulation develops as shown in Figure 11-13 with a surface wind speed of around 10 to 15 knots that may reach about 50 miles inland. The return circulation is around 1,500 to 3,000 feet above the surface.

37. At night the circulation is reversed so that the air movement is from land to sea. This land breeze is generally not as strong as the sea breeze. If the general winds are fairly strong, they will mask the land or sea breeze (Figure 11- 13).

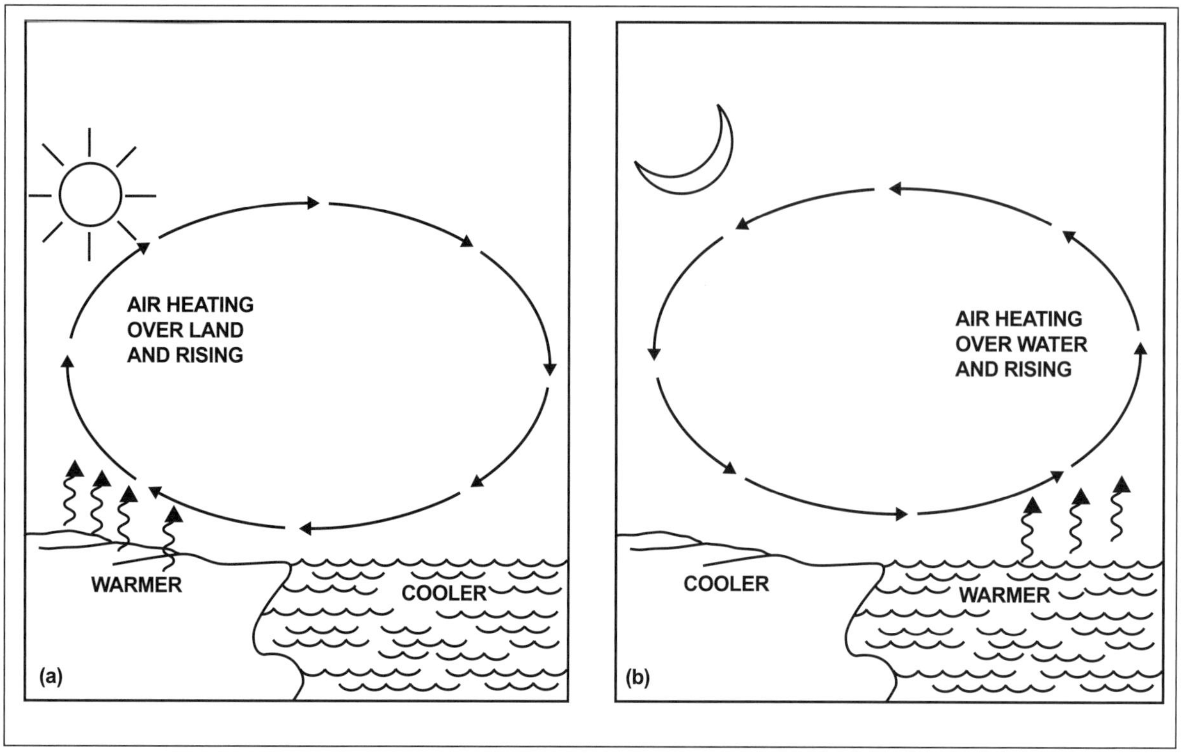

**Figure 11-13 Land and sea breezes**

## Low-Level Nocturnal Jet Stream

38. Fairly frequently, as a nocturnal inversion develops, the wind near the top of the inversion increases to speeds much greater than that indicated by the isobar spacing on a weather map. The wind begins to increase at sundown, reaches its maximum speed two or three hours after midnight and then decreases in the morning as daytime heating destroys the inversion.

39. The level of maximum wind varies from about 700 feet to 2,000 feet above ground. Wind speeds would typically vary from 0-8 knots at ground level to 25-40 knots at the jet maximum, dropping to the gradient winds of 15-30 knots 1,000 feet or so above the maximum wind. In extreme cases, the maximum winds could be in excess of 65 knots with shears of 10 knots per 100 feet below it. There is normally only light turbulence associated with the shear. Low-level jets are one of the causes of hazardous low-level wind shear.

40. The low-level jet is a "sheet" of strong winds some thousands of miles long, hundreds of miles wide and hundreds of feet thick. It generally forms over flat terrain, such as the Prairies, and is most predominant in the summer with moderate to strong south to southwest winds. It can occasionally form elsewhere and with the wind from other directions. Since it is associated with the nocturnal inversion, clear skies are necessary.

**Figure 11-14 The low-level nocturnal jet stream**

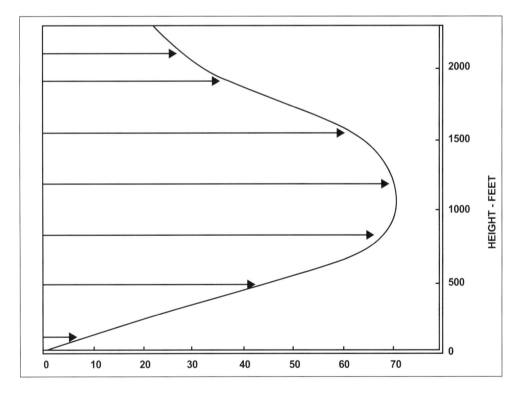

**Figure 11-15 The vertical wind shear with a low-level jet**

# Turbulence

41. There is no task to be done while flying that isn't made more difficult, or at times, impossible, by turbulence. A knowledge of the causes and behaviour of turbulence will be helpful to you to minimize its effects or to avoid it entirely.

42. Turbulence results when the air goes into a random spin motion or develops abrupt updrafts and downdrafts. When the whirls are comparable to the size of an aircraft, they induce chaotic rolls, pitches and yaws. Updrafts, downdrafts and eddies much larger than an aircraft tend to toss it around violently, while eddies smaller than an aircraft are felt as a chop or a tremor. The reaction of the aircraft depends not only on the atmospheric motions but also on aircraft design characteristics and speed. An aircraft with a high wing loading will ride more smoothly through a turbulent area than one with a light loading. A fast flying aircraft will receive more severe jolts than a slow flying one and for this reason, a maximum airspeed for flight in turbulence is laid down for different aircraft types. A large aircraft will react to different size whirls than will a small aircraft. Altitude has an effect because at high altitude the air is less dense and the turbulence tends to soften out. At high altitude, however, the possibility of a turbulence-induced stall is greater, and this can result in loss of aircraft control.

43. The size and strength of the eddies, and the aircraft speed through them translate into different types and intensities of turbulence. Of particular interest is the number of jolts received in a given time, that is, the frequency of the turbulence. Some frequencies have an adverse effect on the crew, others on control of the aircraft and still others on fatigue factors related to the airframe.

44. Figure 11-16 illustrates the effect of different frequencies. The range most strongly affecting airframe fatigue factors is from near .1 to 5 cycles per second. Airsickness tends to occur mainly near .25 cycles per second. Blurred vision is a peculiar effect of turbulence that occurs near 4 to 5 cycles per second. Stability and control problems are greatest from .25 to 3 cycles per second. The eddy wavelengths that affect the crew and aircraft are in the order of tens of metres up to something less than one mile.

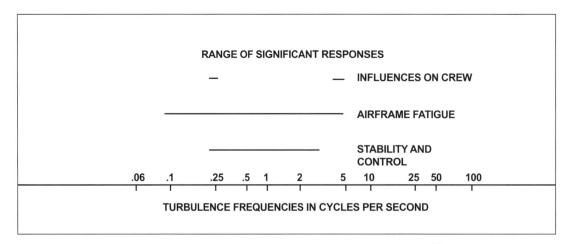

**Figure 11-16 Frequency ranges of turbulence**

45. The intensity of turbulence is forecast and reported using the following criteria:

    a. LIGHT TURBULENCE momentarily causes slight, erratic changes in altitude and/or attitude (pitch, roll, yaw).

    b. LIGHT CHOP is turbulence that causes slight, rapid and somewhat rhythmic bumpiness without appreciable changes in altitude or attitude.

    c. MODERATE TURBULENCE is similar to Light Turbulence but of greater intensity. Changes in altitudes and/or attitude occur but the aircraft remains in positive control at all times. It usually causes variations in indicated airspeed.

    d. MODERATE CHOP is turbulence that is similar to Light Chop but of greater intensity and which causes rapid bumps or jolts without appreciable changes in aircraft altitude or attitude.

    e. SEVERE TURBULENCE causes large, abrupt changes in altitude and/or attitude. It usually causes large variation in indicated airspeed. Aircraft may be momentarily out of control.

    f. EXTREME TURBULENCE tosses the aircraft around violently, making it practically impossible to control. Extreme turbulence may cause structural damage.

46. Some indication of where turbulence occurs has already been provided in this chapter and will now be expanded upon. Turbulence develops through convective activity, wind shear and the movement of air over ground obstacles.

## Convective Turbulence

47. The development of convective cells in unstable air has been described previously. During convection, bubbles or shafts of air rise rapidly through the atmosphere. The cells are some hundreds of feet in diameter and with strong convection the air ascends at the rate of a few thousand feet per minute. The cells are surrounded by slowly descending air. Convection most commonly occurs on a hot summer day over portions of the earth's surface that heat readily. The strength and depth of the convective layer increase as the surface temperature increases. Convection also occurs when a large area of unstable air is forced to rise so that there can be convective cells aloft but none near the earth's surface.

48. The presence of cumulus, towering cumulus, cumulonimbus or altocumulus castellanus indicates that convective currents have developed. If the air is dry, convection may still be present but without the telltale signs of convective cloud. Turbulence will occur in and below the cloud if it has been caused by surface heating and within the cloud if it has developed from general ascent of the air mass. Convective turbulence will cease above the cloud in both cases. The most severe convective turbulence is associated with thunderstorms, which is discussed in detail in a later chapter.

49. Convective turbulence is caused by the aircraft passing out of the slowly descending air into the rapidly rising convective cell. There are also small turbulent eddies caused by shear around the periphery of the cell. The severity of the turbulence depends on how unstable the air is, that is, how fast the air is rising in the convective cell, and the speed of the aircraft. A sailplane flying at 45 knots

may enter a convective cell and climb smoothly in it with only slight ripples being felt around its circumference, whereas a fighter aircraft flying at 600 knots may experience very severe turbulence as it passes rapidly through a series of cells.

**Figure 11-17 Avoiding turbulence by flying above convective cloud**

50. Convection also produces gusts and lulls in the wind speed. The wind shears that result cause rapid fluctuations in airspeed and vertical velocity. These however are very transitory so that any increases in the headwind or tailwind component or up or down drafts are almost immediately followed by decreases. This is totally different than the sustained shears that were described at the beginning of this chapter. Because of the danger of a stall when surface winds are gusty, approach speeds are normally increased slightly.

## Mechanical Turbulence

51. "Mechanical Turbulence" refers to the eddies formed as air flows over various obstructions on the earth's surface. The obstruction may be mountains, hills, buildings or trees, and the resulting air flow can be extremely complicated and chaotic. Turbulence in association with the various local winds that develop in hilly or mountainous terrain and which can be severe, was described in the paragraphs explaining these winds.

52. The rougher the terrain, the more unstable the air and the stronger the windspeed, the greater the turbulence. In unstable air, the eddies tend to grow in size but break up quickly; in stable air they tend not to grow, but dissipate more slowly. If the wind speed is light, eddies tend to remain as rotating pockets near where they form, but if the wind speed is stronger, eddies can be carried away in the general air flow. Figure 11-18 illustrates this situation where turbulence from a hangar is being carried off to a landing area.

53. Turbulence is frequent on the windward side and over the mountain crests if the air crossing the mountains is unstable. It will be less turbulent on the lee side because subsidence stabilizes the air. Very stable air can also develop severe turbulence if the laminar flow that generally occurs with it breaks down into a turbulent flow because of wind speed or abrupt terrain features.

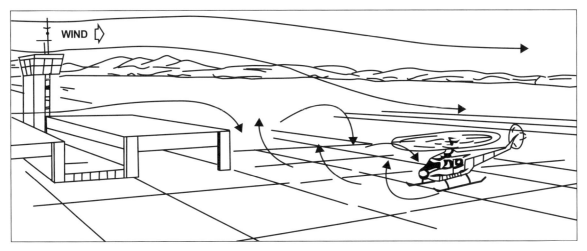

**Figure 11-18 Turbulent eddies being carried away in the general wind**

## Shear Turbulence

54. In this chapter, the boundary layer has been described as a layer where a variety of local winds develop. Frequently there are sharp shear zones between these local winds and the general air flow. Turbulence is sometimes associated with this shear, but not always. Generally, it is not significant but on rare occasions it can be severe. Wind shear is a primary cause of turbulence at high levels and is described in this context later.

## Wake Turbulence

55. An aircraft wing develops lift because the air pressure above it is less than that below it. The air from below the wing streams back and out towards the tip because of this pressure difference and curls over, forming a "Vortex" just inboard of each wing tip. Helicopters as well as fixed wing aircraft produce these vortices.

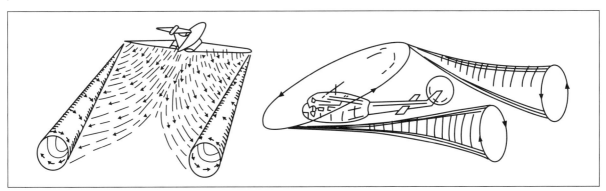

**Figure 11-19 Wing tip vortices**

56. The vortices are 25 to 50 feet in diameter. They stay about three quarters of a wing span apart and sink below the aircraft at 400 or 500 feet per minute, levelling at 700 to 900 feet below the aircraft. They trail back from the aircraft gradually dissipating at distances varying from thousands of feet to several miles depending upon the atmospheric turbulence and the size of the generating aircraft. In general, a wake from an aircraft twice the size of another will extend back twice as far.

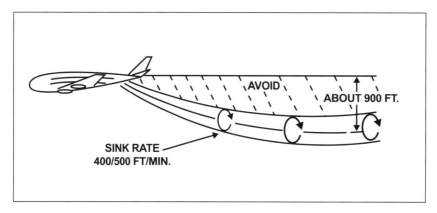

**Figure 11-20 The movement of vortices**

57. There are two scales of turbulence that destroy the vortices. The largest of these is the turbulence that is felt in an aircraft such as convective or mechanical turbulence. It will destroy the vortices within half a minute or so. The other scale is much smaller and occurs anytime that there is a wind. It tends to increase the size of the vortices, which causes the rate of spin to decrease. There will be a substantial weakening of the vortex within a few minutes with this type of turbulence.

58. Conditions where the vortices will persist occur when neither scale of turbulence is present. There is normally an inversion, or at least a stable lapse rate in the stratosphere, so if winds are light enough that little small-scale turbulence is present, vortices may persist for a considerable time. Well shaped contrails extending back for a great distance behind an aircraft may indicate the persistence of vortices, whereas contrails that rapidly deform indicate turbulence that will break up vortices a short distance behind the aircraft. There is generally little attenuation within two minutes of formation of vortices at high level and they sometimes persist for up to fifteen minutes.

59. Figure 11-21 illustrates the hazards of trailing vortices. They include imposed roll, loss of altitude or reduced rate of climb and structural load factors. In addition, severe wake vortices can induce jet engine compressor stall and even flame-out. The extent of the hazards are all dependent on the strength of the vortex. Vortex strength is directly proportional to the aircraft weight and span loading and inversely proportional to the airspeed and time since formation of the vortex. The greatest vortex strength occurs close behind an aircraft that is HEAVY, CLEAN AND SLOW. Small fighter aircraft will have small but very intense vortices. In the case of rotary wing aircraft, the problems created are potentially greater than those caused by conventional aircraft because the helicopter's lower operating speeds produce higher intensity wakes than fixed wing aircraft.

60. If the imposed roll of the wake exceeds the correcting roll available from the ailerons, the airplane will be out of control. In fact the top limit of safe operation is considered to be reached when the correcting roll available is double the imposed roll. If the wing span of the following aircraft is large

enough that the ailerons reach outside the vortex roll, control is usually effective. If the ailerons remain within the vortex roll, control becomes difficult.

61. Pilots should fly at or above a heavy aircraft's flight path, altering course as necessary to avoid the area behind and below it.

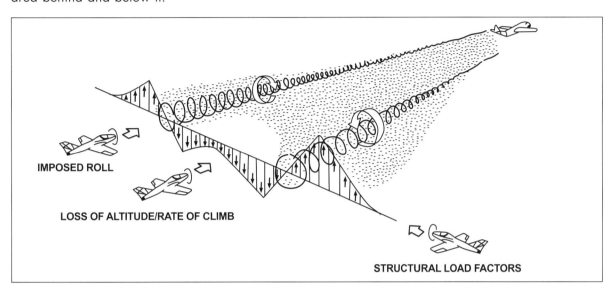

**Figure 11-21 Flight hazards of wake turbulence**

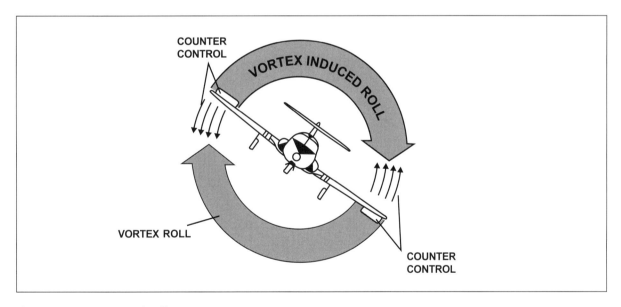

**Figure 11-22 Imposed roll**

## Low-Level Vortices

62. Particular attention must be paid to vortices while taking off or landing. At low levels the vortices sink into ground effect and start moving outward when they have reached about half a wing span height from the ground. This lateral velocity is at speeds up to about 5 knots (Figure 11-23).

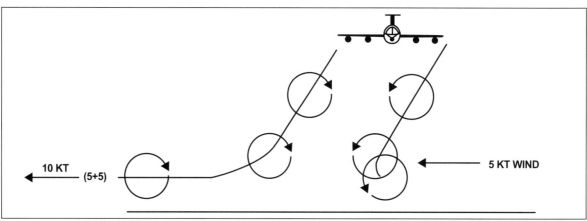

**Figure 11-23 Vortex movement with a crosswind**

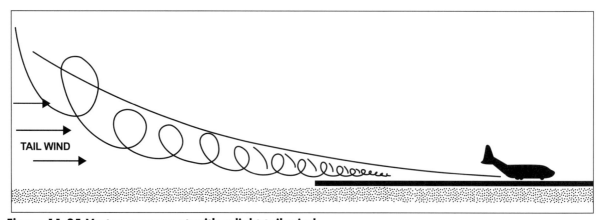

**Figure 11-24 Vortex movement in ground effect**

63. A crosswind component will decrease the lateral movement of the upwind vortex and increase the movement of the downwind vortex (Figure 11-24). This may result in the upwind vortex remaining in the touchdown zone.

64. With a light tail wind, the vortices that have descended below the aircraft during its approach can shift forward into the glide slope area (Figure 11-25).

**Figure 11-25 Vortex movement with a light tail wind**

65. Vortex generation starts on rotation when the nose wheel lifts off the runway and is most severe in that airspace immediately following the point of rotation. It ends when the nose wheel touches down on landing. This leads to the following procedures that are illustrated in Figure 11-26. A following aircraft on departure should take off prior to the point of rotation of a heavy jet and climb above the jet's climb path. An arriving VFR aircraft should approach above the generating aircraft's descent path and touch beyond its touchdown point. An arriving IFR aircraft can normally approach on the glide path since the separation provided by ATC should provide an adequate wake turbulence buffer. Extra caution is required however if the IFR approach is conducted in a light quartering tail wind under conditions where vortices tend to persist (see below). The vortex may be shifted onto the glide path and persist for a considerable time.

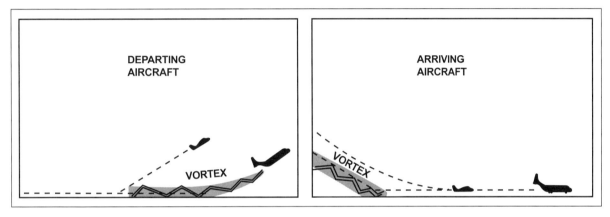

**Figure 11-26 Take-off and landing procedures**

## Persistance of Low-Level Vortices

66. Vortices at low level are initially stronger than they are at high level. Their proximity to the ground and the turbulence near ground level weaken the vortices and hasten their collapse so that they are seldom evident after two or three minutes. Under conditions of extreme stability however, and with a light crosswind, vortices can linger near the runway environment for up to 15 minutes. Such stability will occur under any of the meteorological conditions that will permit a strong surface inversion to develop.

# Summary - Chapter 11

- Due to topography and to differential heating and cooling, the low-level airflow tends to be chaotic and may not be closely related to the isobaric pattern.

- Both wind and abrupt wind shears must be compensated for during flight.

- Abrupt shear causes a change in airspeed which results in a change in the amount of lift being produced.

- An increased performance shear occurs with a rapid increase in head wind or decrease in tail wind.

- With an increased performance shear, airspeed and lift increase.

- A decreased performance shear occurs with a rapid decrease in head wind or an increase in tail wind.

- With a decreased performance shear, airspeed and lift decrease.

- During an approach an increased performance shear will tend to cause an aircraft to rise above the glide path, a decreased performance shear will tend to cause it to sink below the glide path and may cause a stall.

- The surface wind varies diurnally:
     Day - wind increases, becomes gusty, veers.
     Night - wind decreases, becomes smooth, backs.

- A flow of stable air over the ground tends to be laminar, but it can break into turbulence if the speed is great or the terrain features abrupt. A flow of unstable air over hills or mountains will be turbulent.

- Cold air flows downhill, warm air uphill.

- Topography can deflect or cause funnelling of winds.

- An anabatic wind flows up a slope, a katabatic wind flows down a slope.

- Glacier winds and chinook winds are severe downslope winds.

- A land breeze blows from land to sea at night.

- A sea breeze blows from sea to land during the day.

- A low-level jet is a sheet of strong flowing air that blows over plains at night near the top of the nocturnal inversion.

- Turbulence depends on both aircraft and atmospheric factors.

- Turbulence in low level occurs due to convection, surface features (mechanical turbulence) and wind shear.

- Wake turbulence occurs behind aircraft due to the formation of vortices. It is greatest when the generating aircraft is HEAVY, CLEAN AND SLOW.

CHAPTER **11**

# Chapter

# 12

The atmosphere is divided into three layers, each with distinct properties that affect aviation. They are the boundary layer, the troposphere and the stratosphere. Some of these properties can help you and you should learn to use them, others can be a hindrance, and you should learn to avoid these.

# CHAPTER 12

## THE ATMOSPHERE ABOVE THE BOUNDARY LAYER

### The Tropopause

1. You learned earlier that the atmosphere is composed of a lower layer called the troposphere and an upper layer called the stratosphere. The troposphere contains almost all of the moisture in the atmosphere, and the temperature, in general, falls off with height at about 2°C/1,000 feet. The water content of the stratosphere is negligible, and the temperature either does not fall off as fast with height or remains isothermal or even increases with height. The boundary between the troposphere and stratosphere is called the "Tropopause."

### Importance of the Tropopause

2. The tropopause is important because of the following reasons:

   a. It tops the weather layer. Except for cumulonimbus tops that can penetrate several thousand feet into the stratosphere, clouds and weather are confined to the troposphere.

   b. Winds normally increase with height up to the tropopause, and then decrease with height.

   c. Turbulence frequently occurs at the tropopause.

   d. Contrails frequently form and are persistent near the tropopause since it is normally the coldest area. They may dissipate more quickly or not form above or below it.

   e. The performance of an aircraft depends on several factors, one of which is temperature. No temperature advantage can be gained by climbing above the tropopause, and, in fact, temperature may increase markedly above it.

## Variations in Tropopause Height and Temperature

### Height Variations from Equator to Pole

3. Vertical mixing occurs throughout the troposphere due to surface heating. Since more heat is received in equatorial regions than elsewhere, the height of the tropopause is greatest in equatorial regions and lowest in polar areas. Figure 12-1 illustrates the average height of the tropopause from equator to pole. These heights can vary markedly in specific instances, but in general, the warmer the troposphere, the higher the tropopause.

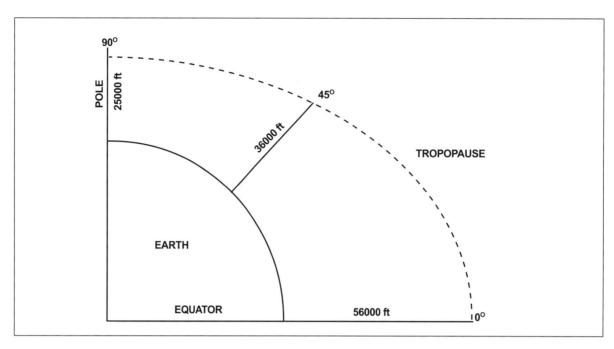

**Figure 12-1 Average tropopause height**

## Height Variations with Air Masses

4.  The troposphere is divided into air masses having different temperatures. These air masses are confined to the troposphere and do not enter the stratosphere. Air masses originating in the hot regions of the world have a high tropopause, those originating in cold areas, a low tropopause. When an air mass moves away from its source region, it carries its tropopause with it. If warm tropical air should move northward over the Canadian Prairies the tropopause may rise to as high as 45,000 feet. With cold arctic air over the Prairies, the tropopause would be around 25,000 feet. This height change of the tropopause occurs rather abruptly at the frontal surface separating the air masses. Figure 12-2 illustrates this. The upper portion of the figure is a vertical cross-section through the line A-B of the frontal wave in the lower portion.

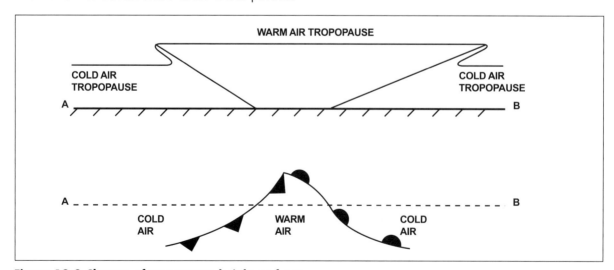

**Figure 12-2 Change of tropopause height at fronts**

B-GA-007-001/PT-D01

5.  As the season changes from summer to winter, the average tropopause height becomes lower because the troposphere becomes colder. Air temperature in the troposphere decreases with height up to the tropopause. This fact means that the temperature at the low tropopause of a cold air mass may be warmer than the temperature at the high tropopause of a warm air mass. That is, the tropopause and stratospheric temperatures are relatively warm over cold tropospheric air masses, and relatively cold over warm tropospheric air masses as shown in Figure 12-3.

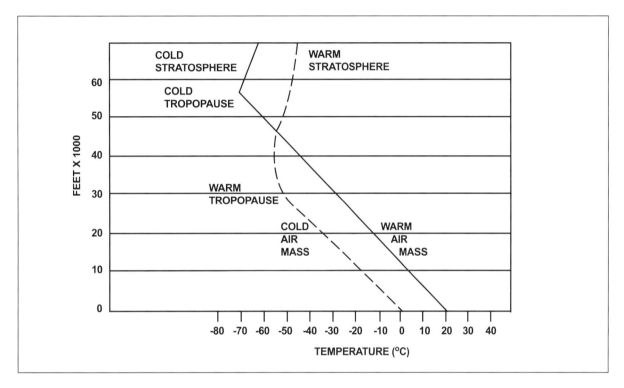

**Figure 12-3 Air mass - stratosphere temperature comparison**

6.  The effects of temperature on aircraft performance is such that efficiency, whether considered in terms of speed or range, is improved with decreased temperature. In high level flight, a temperature advantage will be gained when flying in the cold stratosphere over warm air masses as compared with the warm stratosphere over cold air masses. This temperature advantage is related to the density of the air. It has been found that near 26,000 feet the air density is almost uniform over the globe and from season to season, so that aircraft designed to fly most efficiently near this altitude will show little variation in efficiency from season to season or location on the globe.

## The Arctic Stratosphere

7.  In Chapter 2, the warming of the stratosphere by absorption of solar energy by ozone in the ozonosphere was explained. The ozonosphere is a layer roughly between 6 and 30 miles above the earth's surface. During the winter, the Arctic lies in darkness. As shown in Figure 12-4, the ozonosphere above the Arctic also lies in darkness. Because of this the stratosphere in the polar region becomes very cold during the winter season. The general rule that a relatively warm stratosphere overlies a cold air mass does not apply in arctic regions during the winter.

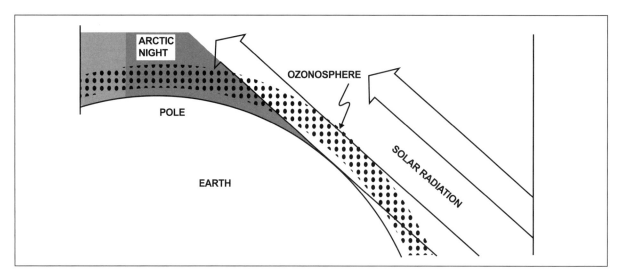

**Figure 12-4 Shading of the Arctic ozonosphere in winter**

## Recognition of the Tropopause

8.  Because of the significance that the tropopause has to flying, the fact that it can frequently be recognized is important.

    a.  A haze layer with a definite top at the tropopause can frequently be seen.

    b.  Anvil tops of thunderstorms spread out at the tropopause. A convective core may burst through the tropopause into the stratosphere, but the anvil will spread out beneath this.

    c.  Winds are normally strongest at the tropopause, decreasing above and below it. This wind shear creates turbulence which is frequently felt as light chop near the tropopause.

    d.  The temperature lapse rate will change at the tropopause. By watching the change in the temperature while the aircraft is climbing, the tropopause can be identified when the temperature ceases to drop.

## Tropospheric and Lower Stratospheric Wind Fields

9.  The wind at about 2,000 or 3,000 feet above the surface, that is, above the friction layer, blows according to the pressure patterns outlined by the surface chart. Wind direction is parallel to the isobars, flowing clockwise around a high and counterclockwise around a low with a speed proportional to the pressure gradient. The flow above this level will now be described. It is important to note that temperature is a major controlling factor as it was with the boundary layer winds.

## Thermal Wind

10. Wind at any level is governed primarily by the pressure distribution at that level. Pressure always decreases with height with the rate of decrease being greater in cold air than in warm air.

11. Consider Figure 12-5 where it is assumed that the MSL pressure is uniform at 1,000 hectopascals. Because there is no pressure gradient at the surface, there is no wind. Cold air lies on the left of the diagram and warm air on the right. The rate of decrease of pressure with height on the left is greater than that on the right so that to reach a pressure of, say, 700 hectopascals, will require a shorter column of cold air than of warmer air. This means that at any particular level, for example, level A, the pressure in the cold air will be less than it will be at the same level in the warm air, and a wind will develop. In the example of Figure 12-5 the pressure at level A in the cold air is 700 hectopascals and in the warm air, 716 hectopascals.

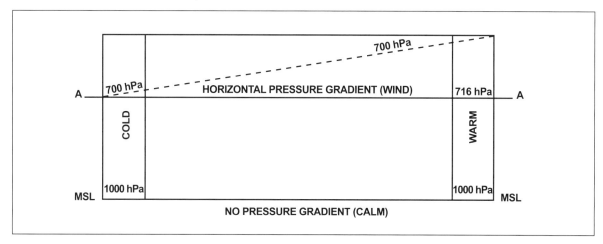

**Figure 12-5 Pressure distribution at a level**

12. Figure 12-6 represents a plan view of Figure 12-5 looking down on level A. Isotherms have been drawn in using broken lines at 5°C intervals for places having the same average temperature in the layer. They indicate cold air on the left and warm air on the right. The pressure differential that has developed at level A has been indicated by drawing in isobars at 4 hectopascal intervals starting at 700 hectopascals. They indicate low pressure to the left and high pressure to the right. It can now be seen that the wind that has developed at level A as represented by the arrow, is parallel to the isotherms with cold air on the left. The stronger the temperature gradient, the greater the pressure difference and the stronger the wind.

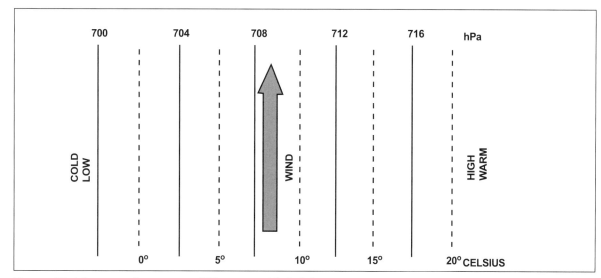

**Figure 12-6 Increased pressure difference with height**

13. If the temperature difference continues up through higher layers, the level-for-level pressure difference will increase and the wind will get continually stronger with altitude. In Figure 12-7 the shaded blocks on the right of the figure indicate how the pressure difference increases with height.

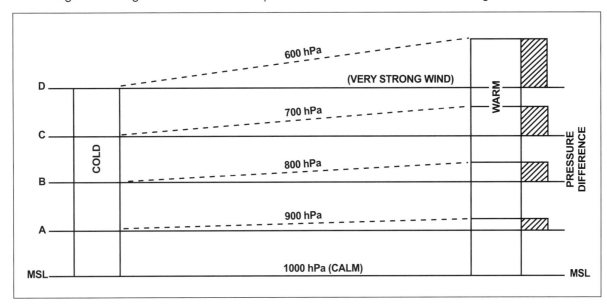

**Figure 12-7 Development of pressure difference with height**

14. There will usually be a wind at the lower level so that surface conditions will not be calm as in Figure 12-5. The wind at any higher level will differ from the lower wind because of the temperature structure. That is, it will be composed of the wind at the lower level plus the wind created by the temperature gradient. This is illustrated in Figure 12-8. The low-level wind is shown as a vector at A, the wind created by the temperature as a vector at B and the resulting upper layer wind as a vector at C. The wind at vector B, which is the difference between the winds at A and C, is called the "Thermal Wind Component" (TWC) since it is due to the temperature structure.

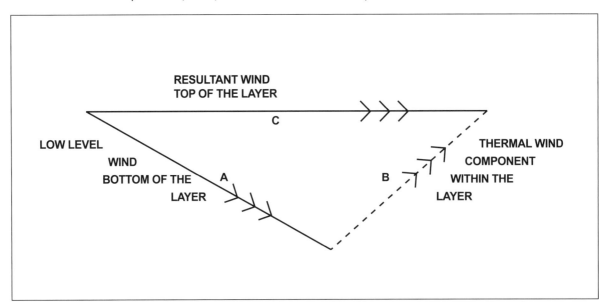

**Figure 12-8 Thermal wind component (TWC)**

15. The TWC is not a real wind, but rather, it is the difference between the actual winds at two different levels. It flows parallel to the isotherms with cold air on the left and with a strength proportional to the temperature gradient (Figure 12-9). A large thermal wind means a large change of actual wind with height, a small thermal wind means little change. In practical terms, this means that if you are flying in an area where there is no temperature gradient, for example, near the centre of an air mass, a change of altitude will produce no significant change in wind. On the other hand, if you are flying where there is a temperature gradient, near the frontal zones at the edge of air masses for example, a change of altitude will produce a change of wind. This change is such that the wind at the higher level will tend to turn parallel to the temperature gradient with colder air to the left looking downwind.

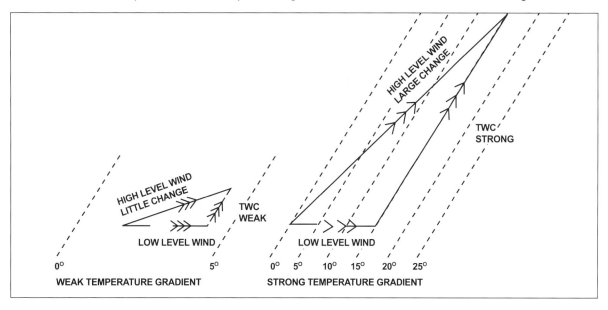

**Figure 12-9 Comparison of thermal wind with temperature gradient**

# Variations of Wind from the Top of the Boundary Layer to the Lower Stratosphere

## Wind in the Troposphere

16. The thermal wind component (TWC) is the major control for high-level winds. Generally speaking, in the troposphere, cold air lies over the polar region and warm air over the tropical region. The thermal wind blows with cold air to the left, therefore the actual wind will swing westerly and increase with height (Figure 12-10(a)).

17. In actual fact, temperature does not decrease uniformly from south to north over the world. As an example, in winter over large land masses such as North America and Eurasia, the air is much cooler than over the adjacent oceans. The resulting temperature structure tends to deflect the winds over the western portions of these land masses towards the southeast and over the eastern portions towards the northeast. Also, cold and warm air masses migrate north and south bringing changes of temperature with them. The resulting upper-level flow, although generally moving towards the east, does meander north and south.

## Wind in the Stratosphere

18. In summer, the temperature regime in the stratosphere is the reverse of what it is in the troposphere (Figure 12-10(b)). The coldest air lies towards the equator and the warmest towards the pole. The stratospheric TWC is therefore easterly. The general westerly flow at the tropopause will be found to gradually weaken with increased height and finally become easterly by around 65,000 feet.

19. In winter, the stratospheric flow is more complicated. The temperature in the Arctic stratosphere gets very cold because it is in the continual Arctic night. A temperature gradient with cold air to the north exists from around 45° north (Figure 12-10(c)). The resulting westerly TWC causes the stratospheric winds to remain westerly. In Arctic regions they become very strong westerly at heights over 65,000 feet.

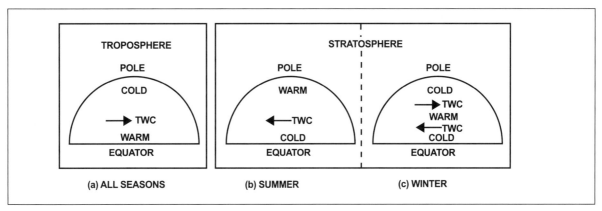

**Figure 12-10 Summer and winter temperature conditions in the troposphere and stratosphere**

20. In summary, if you climb in the stratosphere at our latitudes in summer you will encounter weaker winds and may even reach easterly winds. In the winter, on the other hand, the winds will remain westerly and in far northern regions can become very strong westerly (Figure 12-11).

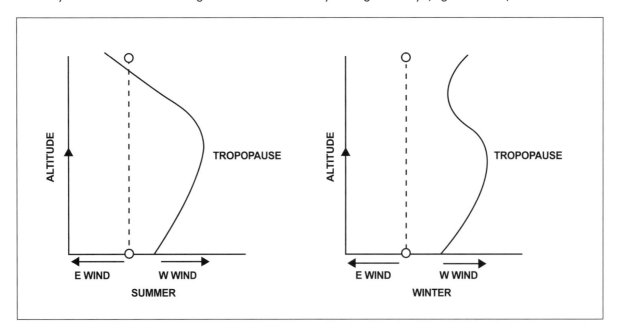

**Figure 12-11 Wind profile**

## Variations of Wind with Height in Relation to Surface Chart Features

21. It is often useful to be able to anticipate wind changes during climb or descent or when changing cruising altitude. Because these changes are related to the temperature distribution, they can frequently be assessed from the surface or upper level charts.

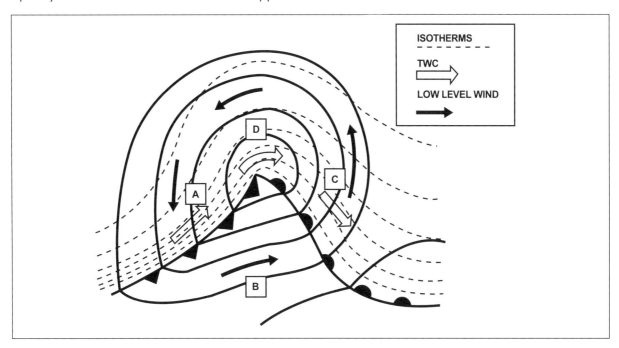

**Figure 12-12 Vertical wind change near a frontal wave**

22. Fronts are zones of strong temperature gradient, so if frontal surfaces lie near the flight route, large variations of wind with height will occur. The temperature change from one air mass to another occurs in the frontal mixing zone. This is illustrated in the diagram of a frontal wave in Figure 12-12. Isotherms representing tropospheric temperatures are shown packed closely together in the frontal zone where the temperature changes from that of the warm air mass to that of the cold air mass. The TWC will blow parallel to these isotherms with cold air to the left at a speed proportional to the temperature gradient.

23. At A, the low-level wind will be of moderate strength from the northwest. The temperature decreases rapidly away from the cold front as shown by the isotherms so the thermal wind will on average be strong southwesterly. On a climb from the surface the wind will back from moderate northwest to strong southwest.

24. In the warm sector at B there is essentially no temperature gradient. The thermal wind will be near zero. The low-level wind will be moderate westerly and there will be little change with height.

25. Ahead of the warm front at C, the low-level wind will be from the south and the TWC will be from the northwest. During climb, winds will on average veer from moderate southerly to strong northwesterly.

26. At D the low-level winds will be moderate easterly. The thermal wind component will be strong westerly so on climb the winds will decrease and then become strong westerly.

27. These variations may occur in a relatively shallow layer from near the surface up to 5,000 or 10,000 feet or they may occur in a thick layer right up to the tropopause.

## Wind Structure Around a Cold Low

28. The occlusion of a frontal wave was described in Chapter 7. In this process, the warm air is forced aloft and moves away from the low. The low is then composed of a whirling mass of air, with a core of cold air in the centre. Because of this temperature structure, pressure falls off with height evenly all around the low resulting in the core being nearly vertical. Also because of the temperature structure, the TWC is counterclockwise around the low the same as the actual wind. The wind speed will therefore increase with height and this can result in very strong easterly winds at high level to the north of the low (Figure 12-13(a)).

29. It is a feature of these lows that they move very slowly so that their associated weather and winds may persist in a region for several days.

## Wind Structure Around a Warm High

30. Cold air is dense and heavy and therefore, particularly in the winter, tends to produce high surface pressure. There is another type of high, however, where the air temperature is very warm and the pressure is high simply because the depth of the atmosphere is exceptionally large over it. This type of high has the warmest air at the centre. The TWC is clockwise around the high, the same as the actual wind, so winds would tend to increase with height but change little in direction. Unlike the cold low, however, the overall wind and temperature pattern is weak, so that strong winds do not develop. Warm highs also move very slowly and persist over an area for a long time. Figure 12-13(b) illustrates a warm high.

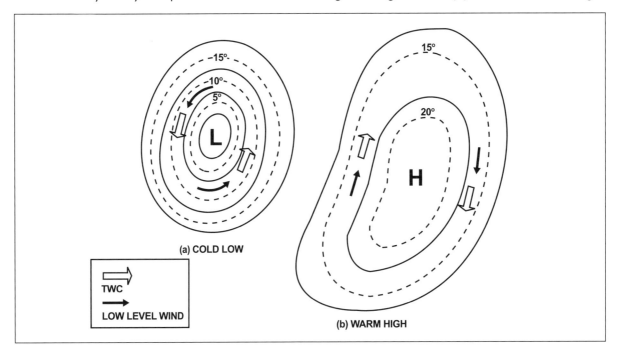

**Figure 12-13 Cold low and warm high**

## Smoothing Out of the Flow with Altitude

31. The action of the TWC is to smooth out the airflow with altitude. The flow near the surface tends to be extremely complicated with the air circulating around highs and lows and changing abruptly at fronts, troughs and ridges. This gradually changes with height into a general west to east flow with occasional north-south meanderings. The main exceptions are the cold lows and warm highs. Generally speaking, aircraft heading corrections will be larger and will be required more frequently to maintain track in the lower levels of the atmosphere than at higher levels.

# Jet Streams

32. You have seen how upper winds are primarily controlled by the temperature structure of the atmosphere below them. Frequently a strong horizontal temperature gradient exists through a deep layer of the atmosphere and, when this occurs, exceptionally strong upper winds called "Jet Streams" develop.

## Definition

33. Jet streams are relatively narrow, rapidly flowing, ribbon-like streams of air embedded within the main airflow. A jet stream may be some thousands of miles long, a few hundred miles wide and a few thousand feet thick. To qualify as a jet stream the speed must be at least 60 knots, although maximum wind speeds can be in excess of 200 knots. The wind speed drops off abruptly above, below and to either side of the jet core. Figure 12-14 illustrates this concept.

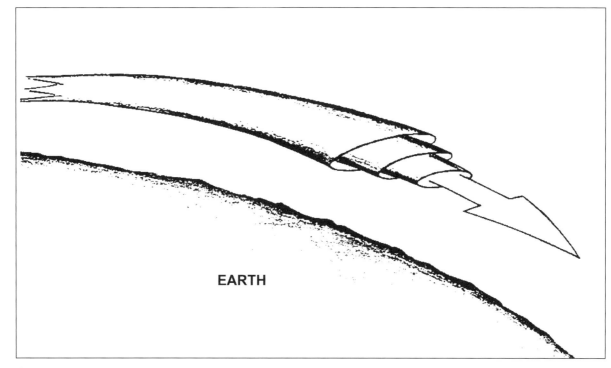

EARTH

**Figure 12-14 Jet stream**

## Frontal Jet Stream

34. If the strong horizontal temperature gradient associated with a front maintains a near-constant direction through a deep layer of the atmosphere the thermal wind component will have a constant direction through successively higher layers. As this component is added consecutively to the actual wind at each level, the wind will increase markedly with height. This is how the very strong winds of frontal jet streams originate. Figure 12-15 shows how the thermal wind component is added to the actual wind up to the tropopause. Fronts do not reach into the stratosphere so the frontal temperature gradient and the resulting thermal wind component cease at the tropopause. In the stratosphere, colder air lies to the south and warmer air to the north causing an easterly thermal wind and a decrease of the real wind with further increase in height.

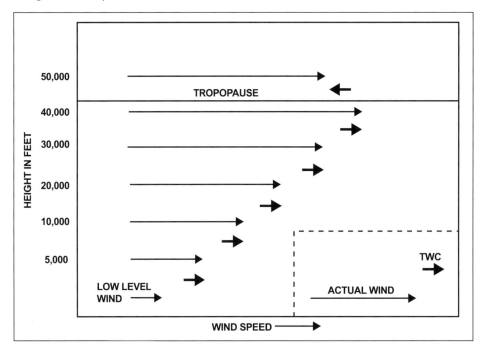

**Figure 12-15 Jet stream development**

35. Figure 12-16 shows a typical jet stream position in relation to a frontal system. It lies to the north of the system where the temperature gradient is greatest and roughly parallels it. If a wave and a low form on the front, they will lie well south of the jet. However, if the wave occludes it will move up under the jet so that the jet cuts over the frontal system near the point of occlusion. The jet will take the name of the frontal system causing it, ie, Polar, Maritime or Arctic.

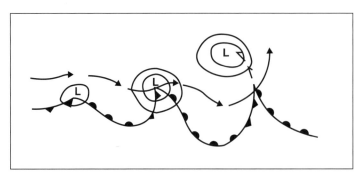

**Figure 12-16 Jet in relation to a frontal system**

36. Figure 12-17 shows the approximate position of the jet core in relation to a frontal surface and to the warm air tropopause and cold air tropopause. The jet lies in the warm air above the frontal surface. Looking downwind, the warm air tropopause lies to the right and above the core, the cold air tropopause to the left and about the same level or a little lower than the core.

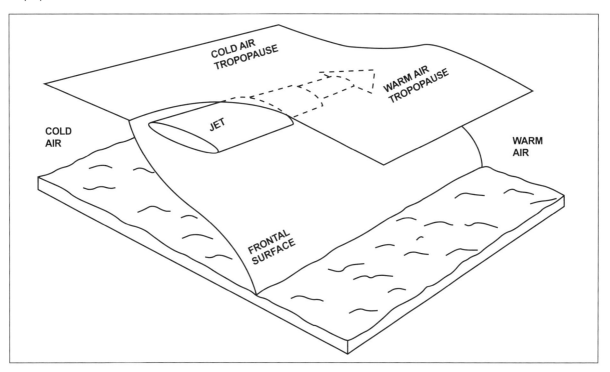

**Figure 12-17 Jet in relation to a frontal surface**

37. Looking downwind, the air in a jet core slowly rotates in a counterclockwise fashion. If the air is moist, the ascending air on the warm air side will cause cirrus cloud to form. This jet stream cirrus takes on very characteristic forms. The strong vertical wind shear may be quite evident from the very wind-swept appearance of the trailing ice crystals of hook shaped cirrus. Sometimes very dense cirrostratus forms with an abrupt poleward edge near the jet core as illustrated in the satellite picture (Figure 12-18). These clouds can at times be recognized in flight, and can help pinpoint the jet core. They will be lying under the warm air tropopause.

38. The tropopause is high over warm air and low over cold air and tends to be higher in summer than in winter. The jet stream is located between the warm air tropopause on one side of the frontal system and the cold air tropopause on the other side and lies at about the same height as the cold air tropopause. The height of the jet stream thus varies according to the front that it is associated with and the season of the year. It also meanders vertically up and down along its length at any particular time. Typical heights of the jet core associated with the different frontal systems are:

   a.  Polar front - 37,000 feet;

   b.  Maritime front - 33,000 feet; and

   c.  Arctic front - 28,000 feet.

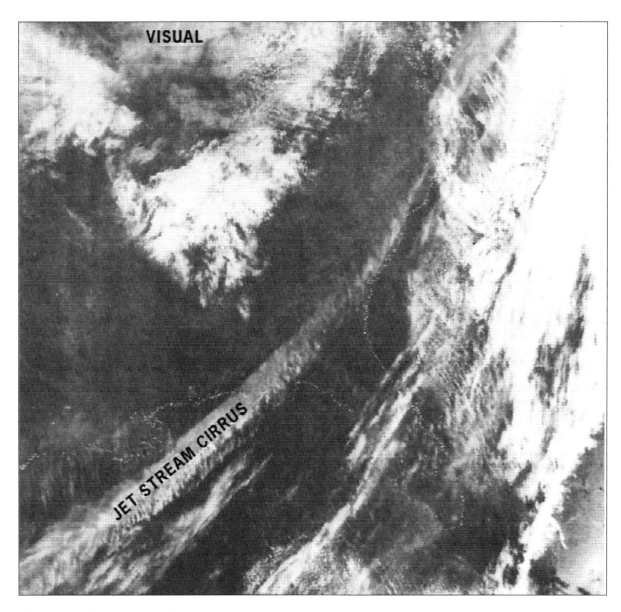

**Figure 12-18 Jet stream cirrus**

## Wind Distribution

39. The very rapid change of wind speed within a short distance of the jet core is a particularly significant feature. Figure 12-19 shows the wind shear on either side of the jet core and above and below it. The vertical shear may vary from around 5 knots per 1,000 feet to extremes of 20 knots per 1,000 feet and is generally close to the same intensity both above and below the core. The horizontal shear on the cold air side of the core (left hand side looking downwind) is stronger than on the warm air side and can vary from around 24 knots per 100 miles to extremes of 100 knots per 100 miles. On the warm air side it is about one third of this. Note particularly that the vertical shear is many times stronger than the horizontal shear. You can utilize this information to adjust your flight to obtain an increased tail wind or a decreased head wind.

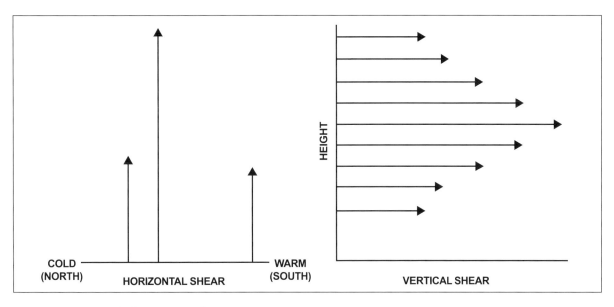

**Figure 12-19 Wind shear around a jet**

40. The speed of the wind along the jet will vary according to the temperature structure underneath it. Where there is a strong temperature gradient aligned so that the thermal wind component is added directly to the wind speed, the overlying section of the jet core will have speeds much greater than the average speed along the axis as a whole. These segments of maximum wind tend to move along the jet core in the direction of the wind. Where the underlying temperature gradient is weak the wind speed in the jet may diminish to the extent that it is no longer strong enough to qualify as a jet stream. Figure 12-20 is an upper level chart that shows two jets lying across the eastern seaboard. Broken lines called "Isotachs" have been drawn joining places having the same wind speed. They outline a 170 knot maximum on the southern jet and a 90 knot maximum on the northern jet. Note particularly that the 90 knot maximum is flowing to the west and is just north of a cold low lying over Newfoundland.

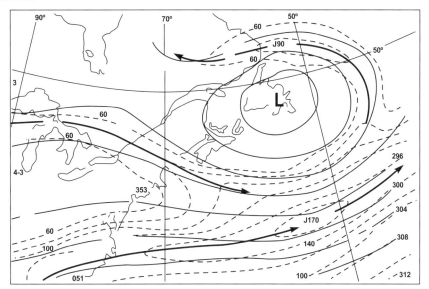

**Figure 12-20 Jet streams**

## Temperature Distribution

41. When the temperature structure of the troposphere and stratosphere was explained, it was shown that the stratospheric temperatures above a warm troposphere are cold and above a cold troposphere are warm. This relationship is shown in Figure 12-21 with a jet core superimposed. Below the core, the normal tropospheric temperature regime prevails with the cold air to the left (north) and warm air to the right (south). Above the core, this is reversed and warm temperatures now lie to the left and cold temperatures to the right. There is little change of temperature through the jet core itself.

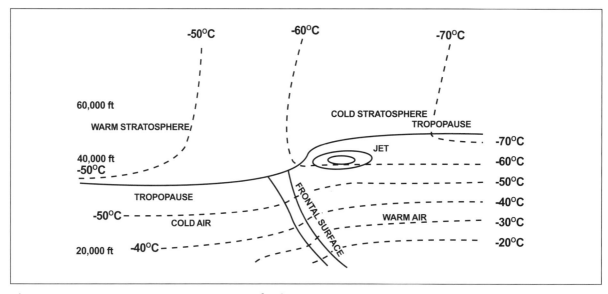

**Figure 12-21 Temperature structure around a jet**

## Seasonal Variations in Latitude and Speed

42. With the tilt of the North Pole away from the sun in winter, frontal zones tend to move further south and tropopause heights tend to become lower in the mid-latitudes of the northern hemisphere. The temperature contrast of air masses is much greater in winter than in summer. This is particularly true over large land masses such as North America and Asia. Thermal winds therefore are much stronger during the winter than during the summer, particularly over large land masses. The result is that jet streams are stronger further south and lower in the winter than in summer, and tend to be at a maximum along the southeastern coasts of North America and Asia where the temperature contrast between land and ocean is the greatest.

## Arctic Stratospheric Jets

43. The development of a strong westerly thermal wind component in the Arctic stratosphere during the winter months was explained in paragraph 20. The thermal wind component was caused as a result of the temperature difference occurring between the shaded and sunlit portions of the ozonosphere in the stratosphere. This thermal wind component is strong enough to create a westerly jet stream called the "Arctic Stratospheric Jet."

44. This jet stream normally occurs at around 70-75° north at heights from 60,000 feet to 80,000 feet. It has, however, extended south as far as 50° north and winds of 100 knots have been found as low as 50,000 feet. Maximum winds at higher levels can exceed 230 knots.

## Subtropical Jet Stream

45. Another jet stream not associated with fronts is the Subtropical Jet Stream. Very strong solar heating in equatorial regions produces a belt of ascending air. This air turns poleward at very high levels. The Coriolis force then turns it to the right into a strong westerly jet (Figure 12-22).

46. The subtropical jet predominates in the winter. It lies near 25° north at around 45,000 feet. If the polar front moves exceptionally far south, the polar front jet stream may merge with the subtropical jet stream.

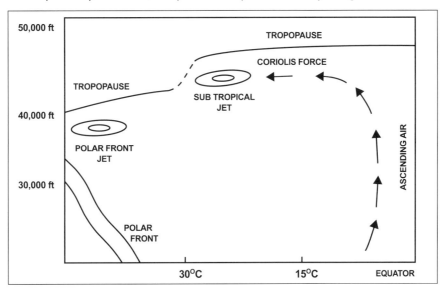

**Figure 12-22 Subtropical jet stream**

## Turbulence

47. Turbulence in the atmosphere above the boundary layer does not occur as frequently as in the boundary layer, but it does occur and can occasionally be hazardous.

48. Convective turbulence has been explained and it can occur anywhere within the troposphere that convection develops. The most serious convective turbulence occurs with thunderstorms which is described later in a chapter on thunderstorms.

49. A form of terrain-induced turbulence associated with mountain waves can occur throughout the troposphere and into the stratosphere. This type of turbulence will be described in Chapter 14, "Mountain Waves."

50. A main cause of turbulence in the atmosphere above the boundary layer is wind shear. Because this type of turbulence can occur in clear air outside of cloud it is called "Clear Air Turbulence."

**Clear Air Turbulence (CAT)**

51. CAT is defined as "all turbulence in the free atmosphere that is of interest to aircraft operations that is not in or adjacent to convective activity." This includes turbulence found in cirrus and other clouds not in or adjacent to visible convective activity and excludes mechanical turbulence.

52. CAT normally causes a rhythmic washboard-like bumpiness that constitutes a nuisance rather than a hazard, but there are some situations where it can cause severe or extreme turbulence. The reporting of the intensity of any turbulence is based on aircraft and occupant reactions. Turbulence is described as:

   a. LIGHT - It causes slight, erratic changes in altitude and/or attitude (pitch, roll, yaw). Airspeed fluctuation will be less than 15 knots.

   b. LIGHT CHOP - It is turbulence that causes slight, rapid and somewhat rhythmic bumpiness without appreciable change in altitude or attitude.

   c. MODERATE - It is similar to light turbulence but of greater intensity. Changes in altitude or attitude occur but the aircraft remains in positive control at all times. It usually causes variations in indicated airspeed of 15 to 25 knots.

   d. MODERATE CHOP - It is turbulence that causes rapid bumps or jolts without appreciable changes in aircraft altitude or attitude

   e. SEVERE - It causes large, abrupt changes in altitude and/or attitude. It usually causes variations in indicated airspeed in excess of 25 knots and the aircraft may be momentarily out of control.

   f. EXTREME - It tosses the aircraft violently, making it practically impossible to control. Extreme turbulence may cause structural damage.

53. CAT occurs in various sized almond-shaped patches that are generally shallow, narrow and elongated with the wind. The occurrence and severity of CAT is influenced by terrain, being very much greater over mountains than over Prairies or oceans. Individual patches of turbulence are shortlived, building up very quickly and lasting from a few minutes to an hour or so. Because of the nature of CAT, precise forecasts are not possible, although areas and levels where turbulent patches are likely to occur, can be forecast (Figure 12-23). On a flight through one of these forecast areas there is at least a 50% chance of encountering turbulence. The patchy nature of CAT makes pilot reports extremely helpful to briefers, forecasters, air traffic controllers and especially to other pilots.

54. Turbulence develops in a shear area when the airflow changes in speed or direction, or both, so abruptly that a smooth airflow is no longer possible and eddies develop. Strong vertical wind shear is the predominant cause of CAT although it can also occur in areas of strong horizontal shear. In order that sufficiently strong shears be present, wind speeds in excess of 50 knots are generally required. For this reason, jet streams are preferred locations for CAT and since jet streams are stronger in winter than in summer, CAT occurs more frequently during winter than during summer.

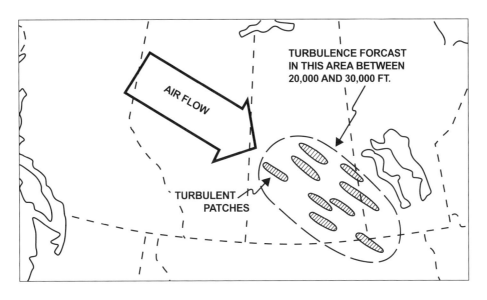

**Figure 12-23 Forecast area of turbulence with embedded turbulent patches**

## Jet Stream Turbulence

55. Not all jet streams have turbulence. However, it has been found that jet streams stronger than 110 knots with vertical shears over 5 knots per 1,000 feet or horizontal shears over 30 knots per 100 miles tend to produce turbulence. Figure 12-24 shows the areas in relation to the jet core where the wind shear is the greatest and the possibility of turbulence most likely. Figure 12-25 shows other areas prone to turbulence. It is frequent on the cold side of the jet stream where it passes around a trough and to the northeast of the surface frontal wave. It can also form in other areas near the jet wherever the shear is great or the direction changes abruptly. The size of the turbulence patches are frequently in the order of 2,000 feet deep, 20 miles wide and 50 miles long, although they can vary markedly from the average.

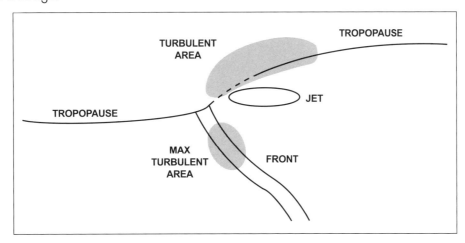

**Figure 12-24 Highest probability of turbulence in relation to jet core**

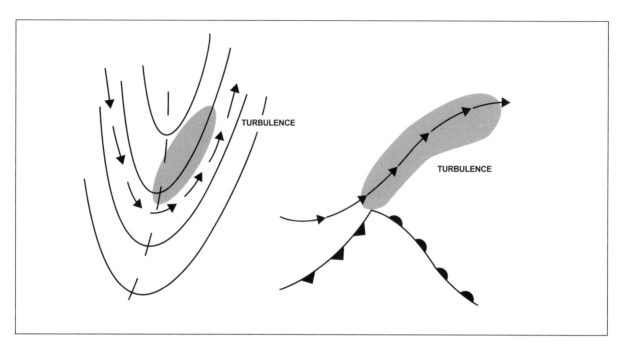

**Figure 12-25 Turbulent areas near jet streams**

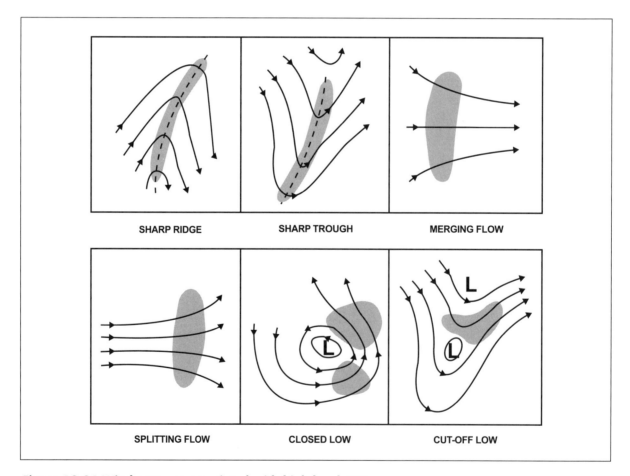

SHARP RIDGE

SHARP TROUGH

MERGING FLOW

SPLITTING FLOW

CLOSED LOW

CUT-OFF LOW

**Figure 12-26 Wind patterns associated with high-level CAT**

## Other Areas of Shear Turbulence

56. In the absence of a jet stream, CAT can occur in wind shears with sharp troughs or ridges, in the circulation around closed and cut-off lows aloft or merging or splitting flows. (See Fig 12-26.)

57. The tropopause is another area where turbulence develops. In this case, it is normally of a light chop variety unless some of the other factors that have been mentioned are also present.

## Turbulence from Evaporation Cooling

58. On occasion, convective cloud develops in mid levels in the troposphere with bases around 12,000 to 15,000 feet from which rain showers fall. If the air beneath the cloud is dry, the rain evaporates before reaching the ground. Rain visible as streaks under the cloud base is called "Virga."

59. The evaporation of the rain cools the air under the cloud, causing it to become denser than the surrounding air. If the air beneath the cloud is extremely dry, such as sometimes occurs over the Prairies, enough evaporation occurs so that the cooled air becomes very dense and accelerates downward. It may strike the ground as a strong, cold gust and spread out from the centre point. Aircraft have had structural damage from encountering this downdraft during flight.

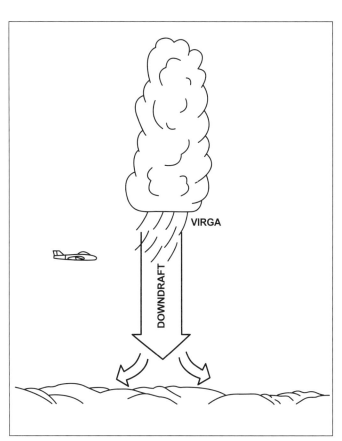

**Figure 12-27 Violent downdraft from evaporation of rain**

## Turbulence Above Storms

60. Turbulence is sometimes found in clear air above rapidly developing thunderstorms. One reason for such turbulence may be that a parcel or column of rapidly ascending air has momentum. If this parcel now invades the wind shear area of a jet stream the air of the jet will be forced to flow around it just as if it were a solid obstruction. The primary conditions for wave formation are satisfied. For example, the wind speed increases with height and the inversion exists aloft near 'mountain top height,' which is, of course, the tropopause.

**Avoiding or Minimizing Encounters with CAT**

61. These additional rules will help in avoiding or minimizing encounters with CAT:

a. Jet streams stronger than 110 knots (at the core) are apt to have areas of significant turbulence near them in the sloping tropopause above the core, in the front below the core, and on the low-pressure side of the core. In these areas, there are frequently strong wind shears.

b. Wind shear and its accompanying clear air turbulence in jet streams are more intense above and in the lee of mountain ranges. For this reason, clear air turbulence should be anticipated whenever the flight path traverses a strong jet stream in the vicinity of mountainous terrain. Flight over areas where the terrain drops abruptly should be avoided. Lenticular cloud may help identify mountain wave conditions but mountain waves are possible in cloud-free conditions.

c. On charts for standard isobaric surfaces, such as 250 hectopascals, if 30-knot isotachs are spaced closer together than 90 nautical miles, there is sufficient horizontal shear for CAT. This area is normally on the poleward (low pressure) side of the jet stream axis, but in unusual cases may occur on the equatorial side.

d. Turbulence is also related to vertical shear. From the FD forecasts, compute the vertical shear in knots per 1,000 feet. If it is greater than about 5 knots per 1,000 feet, turbulence is likely. Since vertical shear is related to a horizontal temperature gradient, the spacing of isotherms on an upper-air chart is significant. If the 5°C isotherms are closer together than two degrees of latitude (120 nautical miles), there is usually sufficient vertical shear for turbulence.

e. Curving jet streams are more apt to have turbulent edges than straight ones, especially jet streams which curve around a deep pressure trough.

f. Wind shift areas associated with pressure troughs are frequently turbulent. The sharpness of the wind shift is the important factor. Also, pressure ridge lines sometimes contain rough air.

g. In an area where significant clear air turbulence has been reported or is forecast, it is suggested that the pilot adjust the airspeed at the recommended turbulent air penetration upon encountering the first ripple, since the intensity of such turbulence may build up rapidly. In areas where moderate or severe CAT is expected, it is desirable to adjust the airspeed prior to the turbulence encounter.

h. If jet stream turbulence is encountered with direct tail winds or head winds, a change in flight level or course should be initiated since these turbulent areas are elongated with the wind, and are shallow and narrow. A turn towards the warm air side of the jet will place the aircraft in more favourable winds. If a turn is not feasible due to airway restrictions, a climb or descent to the next flight level will usually find smoother air.

i. If jet stream turbulence is encountered in a crosswind, it is not so important to change course or flight level since the rough areas are narrow across the wind. However, if it is desired to traverse the clear air turbulence area more quickly, either climb or descend after watching the temperature gauge for a minute or two. If temperature is rising, climb. If temperature is falling, descend. Application of these rules will prevent following the sloping tropopause or frontal surface and thus staying in the turbulent area. (See Figure 12-28) If, however, the temperature remains constant, the flight is probably close to the level of the core, in which case either climb or descend as necessary.

j.  If turbulence is encountered in an abrupt wind shift associated with a sharp pressure trough line, establish a course across the trough rather than parallel to it. A change in flight level is not so likely to alleviate the bumpiness as in jet stream turbulence.

k.  If turbulence is expected because of penetration of a sloping tropopause, watch the temperature gauge. The point of coldest temperature along the flight path will be tropopause penetration. Turbulence will be most pronounced in the temperature change zone on the stratospheric side of the sloping tropopause.

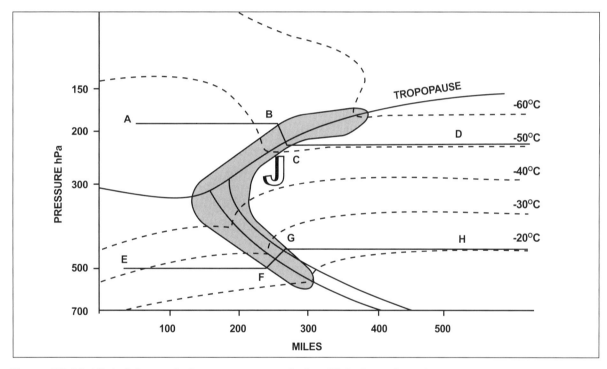

**Figure 12-28 Minimizing turbulence encounter during flight based on the temperature gauge**

62. Figure 12-28 helps explain the 9th rule of thumb for avoiding or minimizing encounters with turbulence. If we start above the tropopause at point A heading towards D we will find that our temperature decreases. If we strike turbulence at point B, a descent will minimize the flight path distance through the turbulence to thousands of feet instead of hundreds of miles. If we reverse the track from D to A turbulence if encountered at C will have increasing temperatures as the tropopause is traversed. This will be the cue to ascend to minimize time in turbulence-prone zones. If we start at point E and fly eastward to H the increasing temperatures as we approach and enter the frontal zone provide the clue that ascent will minimize the time spent in the area of turbulence. If we reverse the track from H to E the pre-frontal cooling will often provide earliest warning so that when turbulence is encountered descent will minimize the time spent in the turbulence-troubled area.

63. Flight at the level of the jet stream will provide no temperature indication but from the general structure of the turbulence (see Fig. 12-28) either climbing or descent will minimize time in turbulence.

## Upper Air Flow Depiction

64. Aviation has a very broad requirement for upper-air wind and temperature information. The prime use of the information is in flight planning. It must be in a form that can be used either for manual or computer methods of flight planning.

65. The need is met in two ways. Charts are provided for different levels in the atmosphere showing the airflow and the temperature over extensive areas of the globe. These include analyses of conditions at several different levels in the atmosphere at a particular time and prognosis of these conditions for different times in the future. This method of providing the information can only be used for manual flight planning.

66. The second method is to provide winds and temperatures at specific locations in a digital format for different heights and at different times in the future. This information can be used for both manual and computer flight planning.

## Upper-Level Analyzed Charts of Standard Pressure Surfaces

67. Temperature, moisture, wind and pressure are measured vertically through the atmosphere at upper-air observing stations around the world by means of instruments called radiosondes. These observations are taken twice a day, at 0000Z (00:00 UTC) and 1200Z (12:00 UTC). The observations, supplemented by information from PIREPS and satellite pictures, are computer-analyzed into charts for several levels in the atmosphere.

68. The levels selected are pressure levels, not geometric heights. This is illustrated in Figure 12-29. Pressure decreases with height at a rate that is dependent on the temperature of the air. In Figure 12-29, the height in feet above sea level where a pressure of 700 hectopascals occurs is illustrated. A general flat plain can be seen at A where a pressure of 700 hectopascals occurs at 10,000 feet. On the right of the diagram, contours are drawn at 200-foot intervals indicating that the 700 hectopascal pressure level rises higher and higher with a final contour drawn at 10,800 feet at B. On the left of the diagram 700 hectopascals occurs at lower and lower levels, down to less than 9,200 feet at C.

69. Figure 12-30, an actual 700 hectopascal (hPa) chart, illustrates how contours appear on an upper-air chart. Wind direction and speed can be estimated directly from the contour lines. These lines can be considered in the same light as isobars on a surface chart. The air flows parallel to them in such a way that, when flying with the wind, the altitude of the pressure surface is lower to the left than to the right. Wind speed on upper level charts is indicated by the spacing between the contours, being the greatest when the contour lines are closest. High altitudes are marked with an H and low altitudes with an L, just as high and low pressure are marked on surface charts. The height of the contours on these charts is marked in decametres.

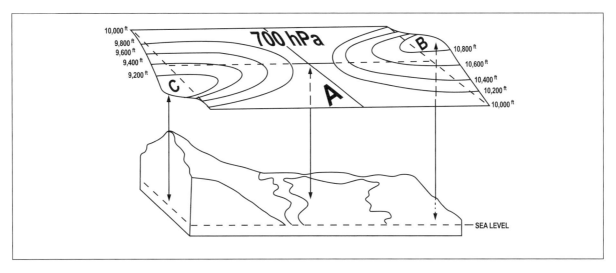

**Figure 12-29 A pressure level**

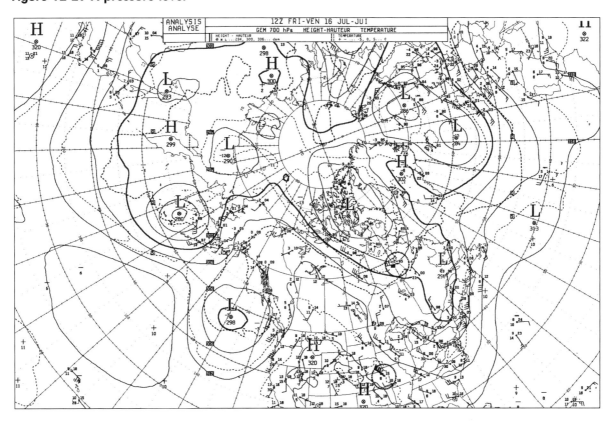

**Figure 12-30 An upper-air chart**

70. The main use of the upper-air analysis charts is in forecasting so the information presented on them is designed for that purpose. Isotherms have been drawn on the chart as in Figure 12-30, but on other charts other features required for forecasting may be analyzed. The standard pressure surfaces, and their approximate heights in feet are:

850 hectopascals - 5,000 feet

700 hectopascals - 10,000 feet

500 hectopascals - 18,000 feet

400 hectopascals - 24,000 feet

300 hectopascals - 30,000 feet
250 hectopascals - 34,000 feet
200 hectopascals - 40,000 feet
100 hectopascals - 53,000 feet

A selection (850, 700, 500, 250 hPa in Canada) of these charts is analyzed and displayed in forecasting and briefing offices.

71. There are advantages to using contour charts of pressure surfaces rather than charts that show the changes of pressure at specific heights. On contour charts, a certain contour spacing produces the same wind at all levels. On an isobaric chart the same spacing between isobars at different heights produces different winds. An aircraft flying a constant altitude indication on its altimeter is flying a pressure surface since an altimeter indicates pressure as height. Because of this the actual height of the aircraft will vary as the height of the pressure surface varies. The changes in height of the aircraft will closely parallel the changes in contour values of the nearest pressure level chart.

## Upper Wind Charts

72. Upper wind charts are specifically designed for aviation to provide information on wind speed and direction and the temperature for various flight levels. The charts are forecasts of conditions as they are computed to be at specific times in the future, generally from 6 to 48 hours ahead. The chart to be used is the one nearest the flight time and the flight level similar charts are provided over most of the world.

73. Figure 12-31 is an upper wind chart that is typically used in aviation. Wind direction and speed are obtained from the wind barbs and are in true degrees and knots. The temperature is given by the number beside the wind barb and is in degrees celsius.

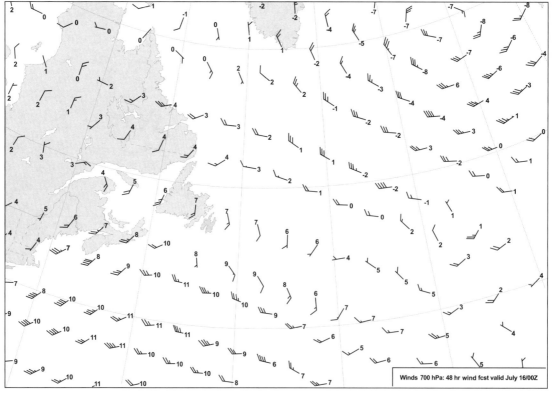

Winds 700 hPa: 48 hr wind fcst valid July 16/00Z

**Figure 12-31 Upper Wind chart**

## Digital Winds and Temperatures

74. Computations similar to those that are used to produce Upper Wind charts can be used to produce forecast digital winds (Figure 12-32). These are provided over Canada from 3,000 feet to 53,000 feet for the locations shown in Figure 12-33 for up to 24 hours ahead. Similar winds are provided over most of the world.

**FDCN01 CWAO 151520**

FCST BASED ON 151200 DATA VALID 151800 FOR USE 17-21

|      | 3000 | 6000     | 9000     | 12000    | 18000    |
|------|------|----------|----------|----------|----------|
| YVR  | 1510 | 2109+12  | 1914+07  | 1720+01  | 1929-12  |
| YYF  | 9900 | 1715+20  | 1924+12  | 1924+04  | 2031-11  |
| YXC  |      | 1406+18  | 1809+12  | 2209+05  | 2425-10  |
| YYC  |      | 1620+15  | 2019+11  | 2510+03  | 2517-11  |
| YQL  |      | 1212+15  | 1617+11  | 2415+05  | 2724-09  |
| YEA  |      | 0205+13  | 3314+09  | 3023+03  | 2922-10  |
| YZP  | 1405 | 1308+09  | 1412+03  | 1415+00  | 1322-13  |
| YZT  | 1605 | 1613+10  | 1517+05  | 1522-02  | 1528-14  |
| YPU  |      | 3406+13  | 0909+06  | 1317-01  | 1617-13  |
| YXS  |      | 1206+15  | 9900+08  | 1606+01  | 1810-13  |
| YYD  |      | 1308+13  | 1307+06  | 1512-01  | 1716-14  |
| YDL  |      | 9900+14  | 1713+05  | 1815-02  | 2017-15  |
| YKA  | 9900 | 1215+18  | 1624+12  | 1723+03  | 1927-12  |
| YJA  |      | 9900+19  | 9900+10  | 2010+01  | 2421-11  |

**FDCN1 KWBC 151407**

**DATA BASED ON 151200Z**

VALID 151800Z   FOR USE 1700-2100Z. TEMPS NEG ABV 24000

| FT   | 24000   | 30000  | 34000  | 39000  | 45000  | 53000  |
|------|---------|--------|--------|--------|--------|--------|
| YVR  | 1944-26 | 204840 | 205351 | 215054 | 213953 | 211554 |
| YYF  | 2139-25 | 214540 | 224750 | 225554 | 224155 | 211957 |
| YXC  | 2434-23 | 244039 | 245548 | 247154 | 245157 | 231456 |
| YYC  | 2531-24 | 264340 | 265349 | 256853 | 244155 | 250956 |
| YQL  | 2634-23 | 264539 | 257346 | 258154 | 255258 | 251557 |
| YEA  | 2838-24 | 285240 | 276448 | 268054 | 265156 | 282157 |
| YZP  | 1323-28 | 151644 | 161353 | 191451 | 191648 | 181050 |
| YZT  | 1533-28 | 153543 | 163551 | 193250 | 202649 | 190851 |
| YPU  | 1722-27 | 182742 | 192951 | 213553 | 202550 | 200952 |
| YXS  | 2217-27 | 232743 | 233151 | 233151 | 211850 | 220752 |
| YYD  | 1809-28 | 231145 | 231552 | 211650 | 191449 | 190850 |
| YDL  | 2115-29 | 211445 | 201051 | 171150 | 201349 | 241349 |
| YKA  | 2134-25 | 224241 | 224650 | 225154 | 223352 | 211355 |

**Figure 12-32 Forecast digital winds**

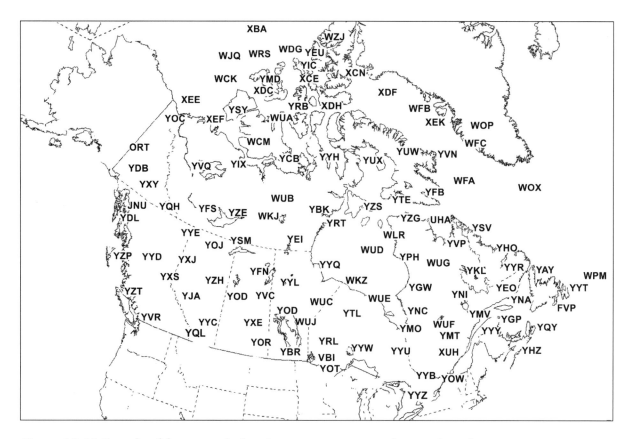

**Figure 12-33 Sample of forecast wind and temperature network over Canada**

## Summary - Chapter 12

- Tropopause heights are high over equatorial areas and warm air masses and low over polar areas and cold air masses.

- The tropopause height change occurs at the frontal surfaces between air masses.

- The tropopause and stratosphere are relatively warm over cold air masses and relatively cold over warm air masses.

- An exception to the above statement is the Arctic stratosphere in winter, which is relatively cold.

- The tropopause can frequently be seen or it can be recognized by the occurrence of chop or by a change in the temperature lapse rate.

- The thermal wind component (TWC) is a fictitious wind that is the vector difference between the wind at two levels in the atmosphere. It blows parallel to the isotherms of the layer with cold air to the left at a speed proportional to the temperature gradient.

- The TWC is the major control of upper winds. Because of it, winds change from complicated erratic winds in the boundary layer and become generally westerly, increasing with height to the tropopause then decreasing in the stratosphere to eventually become easterly. An exception occurs in the north in winter, where the stratospheric winds become strong westerly.

- Variations in the temperature over the earth's surface cause the upper westerly winds to meander north and south.

- An exception to the general westerly flow occurs with cold lows and warm highs. Cold lows can have strong easterly winds at high altitude associated with them.

- Jet streams are relatively narrow, rapidly flowing, ribbon-like streams of air embedded within the main airflow. They are some thousands of miles long, a few hundred miles wide and a few thousand feet thick and have strong wind shear associated with them.

- Frontal jet streams lie north of their parent front in the warm air above the frontal surface with the warm air tropopause above and to the right and the cold air tropopause at the same level or slightly below and to the left. The Polar front jet system is at the highest elevation, the Maritime front jet is next highest with the Arctic front jet being the third highest.

- Two other jet streams are the Arctic stratospheric jet and the subtropical jet.

- Clear air turbulence (CAT) is turbulence in the free atmosphere not related to convective activity. It can occur in cloud and is caused by wind shear.

**12** CHAPTER

- CAT occurs in relatively small sized almond-shaped patches and can develop and dissipate quickly The occurrence of individual patches cannot be forecast but areas where patches are likely to form can be forecast.

- CAT occurs in strong shear areas, especially near jet streams, and near the tropopause.

- Turbulence can also occur under virga from high-based convective clouds.

- The upper-air flow is depicted using analyses of upper-air conditions measured at 0000Z (00:00 UTC) and 1200Z (12:00 UTC), upper wind charts of these conditions and forecast digital winds for specific locations.

# Chapter 13

Next to the airspeed indicator, the altimeter is probably the most important basic instrument in the cockpit. Unfortunately, because of various atmospheric effects, it seldom reads the correct altitude. You must be aware of when the error can become hazardous.

# CHAPTER 13

# METEOROLOGICAL FACTORS IN ALTIMETRY

## The Altimeter

1.  An altimeter is basically an aneroid barometer that indicates the pressure it senses as an altitude in feet. Pressure always decreases with height in the atmosphere and the altimeter is calibrated so that for a specific pressure it senses, it indicates a certain altitude. There are problems in calibration because the relationship between pressure and altitude is not constant. For instance, at sea level pressure, the altimeter should indicate 0 feet, yet sea level pressure constantly varies. The actual heights of different pressures aloft also vary and depend on the temperature of the air and on the surface pressure. For these reasons, the altimeter will seldom indicate the true height of an aircraft.

2.  In order that altimeters can be calibrated so that they will all read the same, they are designed to indicate the correct altitude when conditions of temperature and pressure throughout the atmosphere are those of the ICAO International Standard Atmosphere (ISA). If the atmosphere should happen to be the same as ISA, the altimeter will read correctly except for any non-meteorological instrument errors. If the atmosphere is different than ISA, as it almost always is, there will be an error in the altitude indicated.

3.  Whether the error is important or not depends on the use being made of the altimeter. If the altimeter is being used for the vertical separation of aircraft and all aircraft have the same error, it is not important. If it is being used for landing or for clearing enroute obstacles, the error can be critical.

## The International Standard Atmosphere (ISA)

4.  The International Standard Atmosphere (ISA) constitutes the average conditions of the atmosphere in a temperate climate. The ISA is assumed to have no water vapour in it. Figure 13-1 is a table of the ISA.

5.  A partial list of the values making up the ISA is as follows:

Surface temperature 15°C;
Surface pressure 29.92 inches of mercury (1013.2 hPa);
Lapse rate within the troposphere approximately 2°C/1,000 feet;
Tropopause at approximately 36,000 feet;
Temperature at the tropopause -56.5°C;
Isothermal lapse rate in the stratosphere to approximately 65,000 feet.

Any variation from these conditions may cause an altimeter error.

| ALTITUDE | PRESSURE | | DENSITY | TEMP |
|---|---|---|---|---|
| | HECTOPASCALS (hPa) | INCHES OF Hg | SLUG/FT³ x 10⁻⁴ | °C |
| 0 | 1013.2 | 29.92 | 23,77 | 15.0 |
| 1 000 | 977.2 | 28.86 | 23.08 | 13.0 |
| 2 000 | 942.1 | 27.82 | 22.41 | 11.0 |
| 3 000 | 908.1 | 26.82 | 21.75 | 9.1 |
| 4 000 | 875.1 | 25.84 | 21.11 | 7.1 |
| 5 000 | 843.1 | 24.90 | 20.48 | 5.1 |
| 6 000 | 812.0 | 23.98 | 19.87 | 3.1 |
| 7 000 | 781.8 | 23.09 | 19.27 | 1.1 |
| 8 000 | 752.6 | 22.22 | 18.68 | -0.8 |
| 9 000 | 724.3 | 21.39 | 18.11 | -2.8 |
| 10 000 | 698.8 | 20.58 | 17.55 | -4.8 |
| 11 000 | 670.2 | 19.79 | 17.01 | -6.8 |
| 12 000 | 644.4 | 19.03 | 16.48 | -8.8 |
| 13 000 | 619.4 | 18.29 | 15.96 | -10.8 |
| 14 000 | 595.2 | 17.58 | 15.45 | -12.7 |
| 15 000 | 571.8 | 16.89 | 14.96 | -14.7 |
| 16 000 | 549.2 | 16.22 | 14.48 | -16.7 |
| 17 000 | 527.2 | 15.57 | 14.01 | -18.7 |
| 18 000 | 506.0 | 14.94 | 13.55 | -20.7 |
| 19 000 | 485.5 | 14.34 | 13.10 | -22.6 |
| 20 000 | 465.5 | 13.75 | 12.66 | -24.6 |
| 21 000 | 446.4 | 13.18 | 12.23 | -26.6 |
| 22 000 | 427.9 | 12.64 | 11.82 | -28.6 |
| 23 000 | 410.0 | 12.11 | 11.42 | -30.6 |
| 24 000 | 392.7 | 11.60 | 11.03 | -32.5 |
| 25 000 | 376.0 | 11.10 | 10.65 | -34.5 |
| 26 000 | 359.9 | 10.63 | 10.28 | -36.5 |
| 27 000 | 344.3 | 10.17 | 9.92 | -38.5 |
| 28 000 | 329.3 | 9.72 | 9.57 | -40.5 |
| 29 000 | 314.8 | 9.30 | 9.23 | -42.5 |
| 30 000 | 300.9 | 8.89 | 8.89 | -44.4 |
| 31 000 | 287.4 | 8.49 | 8.56 | -46.4 |
| 32 000 | 274.5 | 8.11 | 8.24 | -48.5 |
| 33 000 | 262.0 | 7.74 | 7.94 | -50.4 |
| 34 000 | 250.0 | 7.38 | 7.65 | -52.4 |
| 35 000 | 238.4 | 7.04 | 7.37 | -54.3 |
| 36 000 | 227.3 | 6.71 | 7.08 | -56.3 |
| 37 000 | 216.6 | 6.40 | 6.77 | -56.5 |
| 38 000 | 206.5 | 6.10 | 6.45 | |
| 39 000 | 196.8 | 5.81 | 6.14 | |
| 40 000 | 187.5 | 5.54 | 5.85 | |
| 41 000 | 178.7 | 5.28 | 5.58 | |
| 42 000 | 170.4 | 5.04 | 5.32 | |
| 43 000 | 162.4 | 4.79 | 5.07 | |
| 44 000 | 154.7 | 4.57 | 4.83 | |
| 45 000 | 147.5 | 4.35 | 4.60 | CONSTANT |
| 46 000 | 140.6 | 4.15 | 4.38 | AT |
| 47 000 | 134.0 | 3.96 | 4.18 | |
| 48 000 | 127.7 | 3.77 | 3.99 | -56.5°C |
| 49 000 | 121.7 | 3.59 | 3.80 | |
| 50 000 | 116.0 | 3.42 | 3.62 | TO 65 000 ft. |
| 51 000 | 110.5 | 3.26 | 3.45 | |
| 52 000 | 105.3 | 3.11 | 3.29 | |
| 53 000 | 100.4 | 2.96 | 3.14 | |
| 54 000 | 95.7 | 2.83 | 2.99 | |
| 55 000 | 91.2 | 2.69 | 2.85 | |
| 56 000 | 86.9 | 2.57 | 2.72 | |
| 57 000 | 82.8 | 2.45 | 2.59 | |
| 58 000 | 79.0 | 2.33 | 2.47 | |
| 59 000 | 75.2 | 2.22 | 2.35 | |
| 60 000 | 71.7 | 2.12 | 2.24 | |

**Figure 13-1 International Standard Atmosphere**

6. In the lower 5,000 feet, the pressure changes approximately 1 inch of mercury for every 1,000 feet of altitude (1 hectopascal for every 30 feet). At 20,000 feet, this becomes approximately 0.5 of an inch for every 1,000 feet and at 40,000 feet, 0.3 of an inch. To calibrate an altimeter it is therefore necessary to equate a 1,000-foot height change to about a 1-inch pressure change at low levels but to only about a 0.3-inch pressure change at high levels.

7. The altitude designated for each pressure in the International Standard Atmosphere is called the "Pressure Altitude."

8. As well as temperature, pressure and height relationships, the International Standard Atmosphere defines many of the physical parameters required for aerodynamics, missile design and ballistics. Density is of particular importance for aircraft performance and design; specific values for it are computed throughout the ISA. The density varies directly with the pressure and inversely with the temperature. If the temperature in the atmosphere is warmer than ISA at a pressure level, the density will be less than ISA and aircraft performance and indicated airspeed will decrease. If the temperature in the atmosphere is colder than ISA at a pressure level, the density will be greater than ISA and performance and indicated airspeed will increase.

9. Figure 13-3 illustrates how pressure and density change with height and how this change varies in air masses that are colder and warmer than ISA. The chart shows that variations in temperature from ISA will affect aircraft performance more at low levels than at high levels and will affect altimeter errors more at high levels than at low levels. One very significant effect of the low-level density variation is the marked decrease in take-off performance of aircraft in hot weather.

## Surface Pressure Error

10. Consider a simple altimeter designed to work only under ISA conditions. In Figure 13-2(a), an aircraft has climbed from 0 feet (29.92 inches) through an ISA atmosphere to a pressure of 28.86 inches. As indicated in Figure 13-2 this pressure occurs at a pressure altitude of 1,000 feet in the International Standard Atmosphere. The altimeter will read 1,000 feet and the aircraft will actually be at 1,000 feet.

**13** CHAPTER

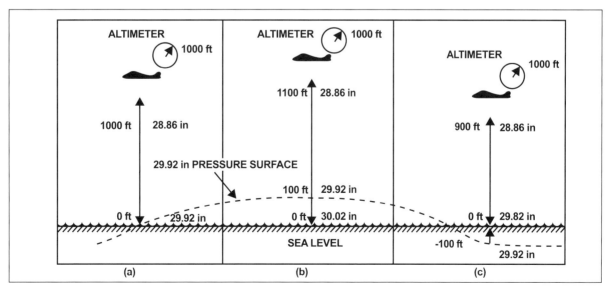

**Figure 13-2 Surface pressure error**

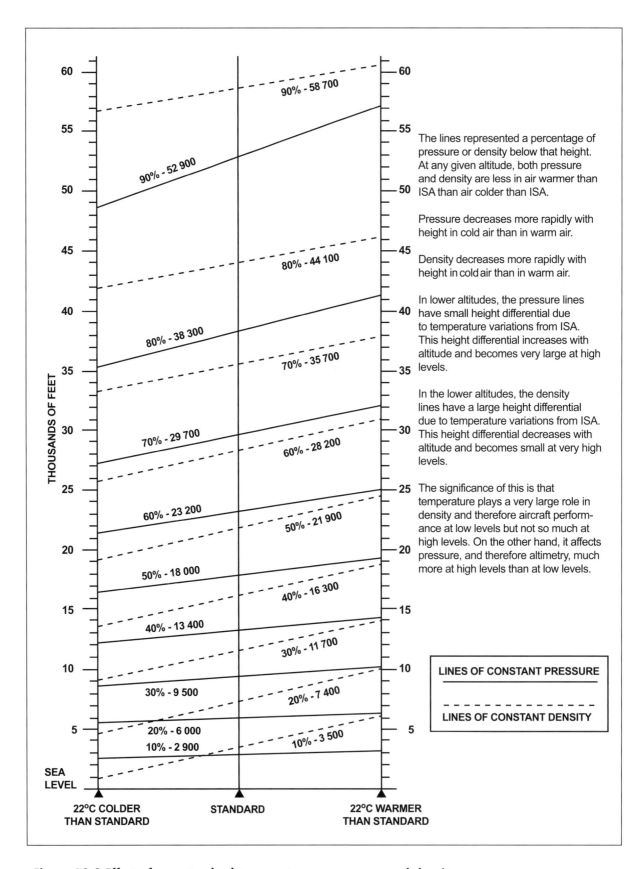

The lines represented a percentage of pressure or density below that height. At any given altitude, both pressure and density are less in air warmer than ISA than air colder than ISA.

Pressure decreases more rapidly with height in cold air than in warm air.

Density decreases more rapidly with height in cold air than in warm air.

In lower altitudes, the pressure lines have small height differential due to temperature variations from ISA. This height differential increases with altitude and becomes very large at high levels.

In the lower altitudes, the density lines have a large height differential due to temperature variations from ISA. This height differential decreases with altitude and becomes small at very high levels.

The significance of this is that temperature plays a very large role in density and therefore aircraft performance at low levels but not so much at high levels. On the other hand, it affects pressure, and therefore altimetry, much more at high levels than at low levels.

LINES OF CONSTANT PRESSURE

LINES OF CONSTANT DENSITY

**Figure 13-3 Effect of non-standard temperature on pressure and density**

11. In Figure 13-2(b), the surface pressure has increased to 30.02 inches. Since 0.1 inch approximates 100 feet, the 29.92-inch pressure level will be 100 feet above the surface. When the aircraft is on the surface, the altimeter will read -100 feet. On climbing to a pressure of 28.86 inches the altimeter will again read 1,000 feet, however the aircraft will be at 1,100 feet above the surface. That is, with a surface pressure higher than ISA the aircraft is higher than the height indicated on this altimeter. The error is the same for all altitudes and is close to 100 feet for every 0.1 inch of surface pressure difference above 29.92 inches.

12. In Figure 13-2(c) the surface pressure has dropped to 29.82 inches. The 29.92-inch pressure surface will now be 100 feet below the surface. When the aircraft is on the surface the altimeter will read 100 feet. On climbing to a pressure of 28.86 inches, the altimeter will again read 1,000 feet but the aircraft will only be 900 feet above the surface. In this case, with a surface pressure lower than ISA, the aircraft is lower than the height indicated on the altimeter by about 100 feet for every 0.1-inch difference from 29.92 inches.

13. In each case, it can be seen that this simple altimeter is indicating an altitude of 1,000 feet above the 29.92-inch pressure level.

## Altimeter Setting

14. Changes occurring in surface pressure are compensated for in an altimeter by means of the "Pressure Adjustment Knob." Figure 13-4 illustrates a typical altimeter and indicates in a simplified way how the pressure adjustment knob moves the aneroid mechanism and adjusts the altitude indicating needle. The amount of the adjustment is indicated by the altimeter subscale. As indicated previously, in the lower 5,000 feet an increase of 0.1 inch on the subscale will increase the altitude indicated by about 100 feet and a decrease of 0.1 inch will decrease it by about 100 feet. The pressure adjustment knob simply realigns the pressure-altitude relationship to set zero feet against some pressure other than 29.92 inches.

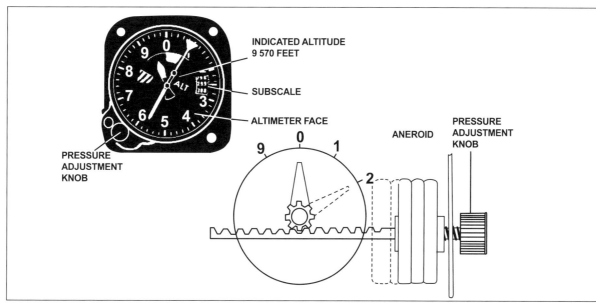

**Figure 13-4 Typical aneroid altimeter**

15. In Figure 13-2, the 29.92-inch pressure surface is shown as a broken line. In Figure 13-2(a) conditions are ISA and the aircraft is at the altitude indicated on the altimeter. If the pilot now flies into the conditions shown in 13-2(b) holding 1,000 feet on the altimeter, he/she flies 1,000 feet above the 29.92-inch pressure surface ending up with an actual altitude of 1,100 feet. If the pilot now adjusts the altimeter to 30.02 inches this becomes the 0 height pressure and the indicated altitude will change to 1,100 feet. On the other hand, if he/she flies into the conditions indicated in 13-2(c) and adjusts the altimeter to 29.82 inches, the reading will change to 900 feet. The pressure set on the subscale becomes the zero altitude pressure and the ISA pressure-height relationships are realigned accordingly.

16. Consider an aerodrome at 4,000 feet ASL, as shown in Figure 13-5. The requirement is to provide a pressure that when set on the subscale of the altimeter will cause it to read 4,000 feet when the aircraft is on the aerodrome. This is done by measuring the station pressure, 24.90 inches in the example, and going from this pressure to the pressure 4,000 feet below it in the standard atmosphere, 28.86 inches in this case. If 28.86 inches is set on the altimeter of an aircraft at any altitude, it will read 4,000 feet when the aircraft alights on the aerodrome. This pressure is called the "Altimeter Setting."

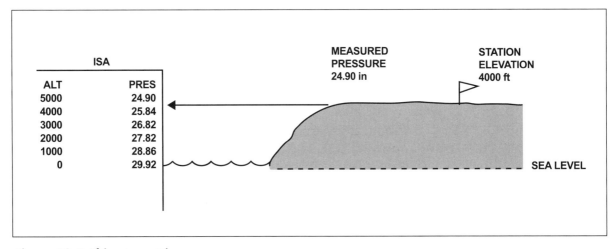

**Figure 13-5 Altimeter setting**

17. The "Altimeter Setting" is the pressure that when set on the altimeter will cause the altimeter to read station elevation when the aircraft is on the ground at the aerodrome. It is used throughout the altimeter setting region of Canada. The altitude obtained using the altimeter setting is called the "Indicated Altitude."

18. The altimeter setting is used during take-off and landing and in flight in the low-level airspace. On the other hand, in North America, 29.92 inches is set on the altimeter for flight in the highlevel airspace and in the Standard Pressure Region of Northern Canada at all altitudes. The resulting altitude then shown on the altimeter is the "Pressure Altitude." For each pressure sensed, the altimeter will display the appropriate ISA altitude.

19. Using standard pressure of 29.92 inches, the altimeter acts like the simple instrument described in paragraphs 10 to 13 so that the actual height of an aircraft holding a steady altimeter reading varies with the surface pressure. The altimeter setting, on the other hand, corrects for the surface pressure variation.

## Terrain Clearance in the Standard Pressure Region

20. You should be very alert for surface pressure errors when flying near minimum obstacle clearance altitudes in the Standard Pressure Region (Figure 13-6). By studying the surface weather chart, you can assess whether your altimeter will be overreading or underreading and approximately by how much. If the MSL pressure as outlined on the surface chart is below the ISA zero feet altitude pressure of 1013.2 hectopascals, your aircraft will be lower than indicated by about 30 feet for each hectopascal. For instance, if a low with a central pressure of 972 hectopascals lay on your route, your altimeter would overread by (1,013-972) x 30 = 1,230 feet as you overflew the low pressure. This error occurs at all altitudes.

21. A convenient way to remember this concept is the phrase, "from high to low, look out below." When flying from high to low surface pressure with a constant altimeter setting, the aircraft will become lower than the altitude indicated by the altimeter. The amount is about 30 feet for every hectopascal surface pressure difference or 100 feet for every 0.1 inch of mercury difference.

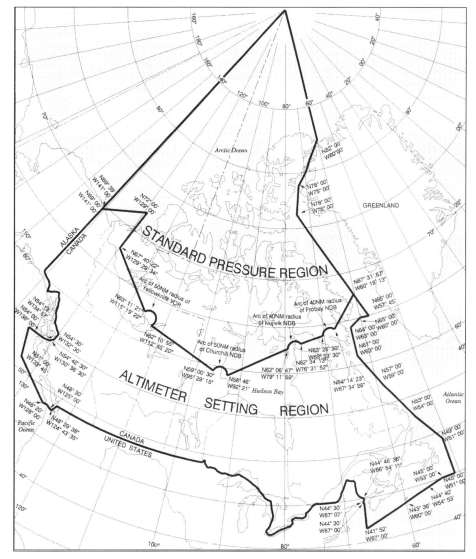

**Figure 13-6 Standard Pressure Region**

## Drift and Altimeter Error

22. Figure 13-7 illustrates an aircraft flying with 18,000 feet on the altimeter set at 29.92 inches. The aircraft is flying on a 500 hectopascal pressure surface because 18,000 feet is 500 hectopascals in the International Standard Atmosphere. Air circulates counterclockwise around centres of lowest heights of pressure surfaces and clockwise around centres of highest heights. In the illustration, the aircraft has been drifting to the right and since it is flying a constant pressure surface its actual altitude is decreasing. Were the aircraft to fly in the opposite direction, it would drift to the left and its actual altitude would be increasing. The wind speed is proportional to the slope of the pressure surface, so the greater the drift, the more rapidly the aircraft is changing altitude. A method of navigation called "Pressure Pattern Flying" utilizes this change in altitude to compute the drift.

## Altitude Variation and Altimeter Settings

23. Figure 13-8 shows an aircraft flying at an indicated altitude of 5,000 feet towards higher surface pressure. The altimeter setting is being set on the altimeter as it progresses along the route. In this case, the drift will be to the left and the actual aircraft altitude will be increasing. When each new altimeter setting is placed on the altimeter it will indicate that the aircraft has climbed above 5,000 feet and the aircraft will have to descend to regain its correct indicated altitude (IA).

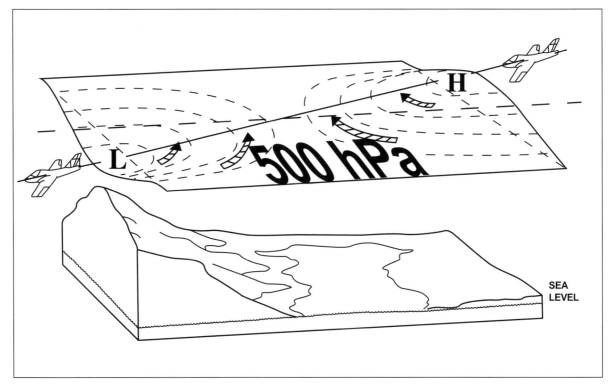

**Figure 13-7 Drift and altimeter error**

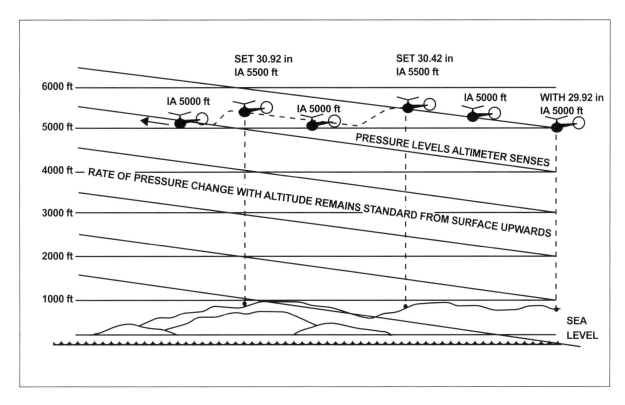

**Figure 13-8 Altitude variations and altimeter settings**

24. From this it can be seen that a too low subscale setting on the altimeter will mean that the aircraft will be higher than indicated, and a too high subscale setting, that the aircraft will be lower than indicated. For example, should a pilot, through error, set 29.92 inches on the altimeter rather than 28.82 inches when descending to land, the aircraft will be 1,000 feet lower than the altimeter is indicating.

## Temperature Errors

25. The effect of nonstandard temperatures on the height of pressure surfaces aloft was shown in Figure 13-3. Pressure surfaces are lower in cold air than in warm air and the difference increases progressively with altitude. The amount of the error is proportional to the average temperature throughout the air column between the aircraft and the ground and diminishes to zero as the aircraft approaches the ground at the point where the pressure was measured. Figure 13-9 compares an atmosphere 15°C colder than ISA with an ISA atmosphere.

26. Consider an aircraft flying at an indicated altitude of 10,000 feet (700 hPa). As indicated in the left hand diagram of Figure 13-9 under ISA conditions, the surface temperature would be 15°C and the 700 hPa temperature -5°C. The centre diagram illustrates a situation where the air column is 15°C colder than ISA. The surface temperature is 0°C, the 700 hPa temperature is -20°C and the lapse rate is identical to the ISA lapse rate. The 700 hPa surface is considerably lower than under ISA conditions. In actual fact it is about 9,400 feet.

27. The right hand diagram of Figure 13-9 also illustrates a situation where the air column averages 15°C colder than ISA. Note particularly that the temperature at aircraft altitude (700 hPa) in this case is identical to the ISA temperature, but a surface inversion brings the average temperature of the column down to 15°C below ISA. For this reason the 700 hPa surface is again about 9,400 feet.

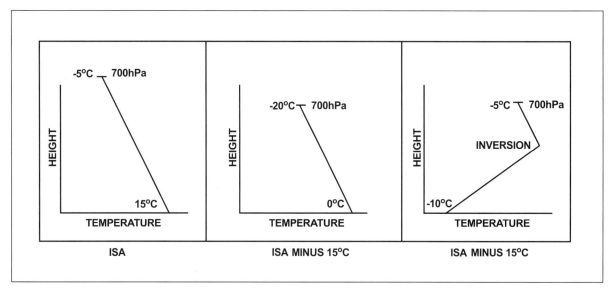

**Figure 13-9 Average air column temperature**

28. If the average temperature of the air column beneath the aircraft is warmer than ISA, the pressure surfaces will be higher than with ISA.

29. The amount of the error caused by temperature can be calculated using the outside air temperature at the aircraft altitude provided that the lapse rate below the aircraft is very close to the ISA lapse rate. The correction can be made using a flight computer or can be approximated by using a factor of four feet per degree from ISA per 1,000 feet. It is the height of the column of air between the aerodrome and the aircraft that is required and not the height between sea level and the aircraft. As an example, an aircraft with an indicated altitude of 9,000 feet ASL is using an altimeter setting from an aerodrome at 1,000 feet and has an OAT 30°C below ISA. The column of air from the aerodrome to the aircraft is 8,000 feet deep so the error is 4 x 30 x 8 = 960 feet. Note that this assumes an ISA lapse rate. Since the air is colder than ISA, the aircraft will be lower than the indicated altitude.

30. There is a strong surface inversion in winter throughout much of Canada so the average temperature below the aircraft is frequently much colder than that obtained using ambient air temperature at altitude. Since there is no way of knowing what the temperature structure actually is, the error found using flight level temperature in these conditions should be increased by 50% to add a safety margin. In the given example, if a surface inversion existed, it should be assumed that the error is approximately 1,440 feet.

31. The indicated altitude corrected for the flight level temperature is called the "True Altitude." You must remember that it is "true" only if a standard lapse rate exists below the aircraft.

## Combined Errors

32. Pressure and temperature errors can combine to give an exceedingly large total error. Consider the situation where an aircraft with 29.92 inches set on the altimeter flies over a 972 hectopascal surface low. The pressure error is 1,230 feet (paragraph 20). If this aircraft were flying at 10,000 feet pressure altitude, its height would be 8,770 feet. If the air temperature were 30°C colder than ISA, and rounding 8,770 feet to 9,000 feet, the temperature error would be 4 x 30 x 9 = 1,080 feet. This gives a combined error of 2,310 feet. Even a 2,000-foot clearance over mountains would NOT suffice.

33. There are occasions when placing an altimeter setting on the altimeter worsens the situation. Figure 13-10 illustrates an aircraft flying an indicated altitude of 10,000 feet (700 hPa) towards a surface high pressure area composed of air much colder than ISA. Because the surface high is so cold, the 700 hectopascal surface is lower than ISA even though the surface pressure is high. In the figure, the aircraft passes over point A where the atmosphere is ISA. It is flying 10,000 feet indicated altitude with 29.92 inches set on the altimeter. Because of the ISA conditions, there is no altimeter error. As the aircraft progresses along the route, the 700-hPa surface on which it is flying steadily lowers because of the cold temperatures, so the aircraft's actual altitude lowers. At point B, a new altimeter setting of 30.62 inches is obtained and set on the altimeter. This is 0.7 inch higher than it was previously so the aircraft must descend 700 feet to regain an indicated altitude of 10,000 feet. Placing the altimeter setting on the altimeter has caused the aircraft to fly lower than it otherwise would have.

34. This type of situation occurs fairly frequently over Western Canada when a large Arctic high builds up to the east of the Rockies. A study of an upper level chart with a height near the flight level will indicate if the pressure surface is lower than ISA and by how much. If it is lower, extra height should be allowed for a minimum obstacle clearance altitude.

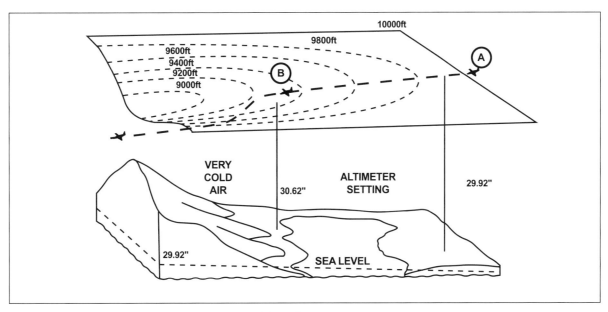

**Figure 13-10 Altitude error with surface cold high pressure**

**Density Altitude**

35. "Density Altitude" is sometimes required to compute aircraft performance. Density altitude is the pressure altitude (subscale set at 29.92 inches) corrected for the difference in temperature from ISA at the level of concern. This correction can be obtained using a flight computer or can be approximated by adding or subtracting 120 feet to the pressure altitude for every degree Celsius that the ambient temperature differs from ISA. For example, a pressure altitude of 5,000 feet with a temperature of 20°C (15°C above ISA) gives a density altitude of 6,800 feet. (120 x 15) + 5,000.

36. The ISA contains no water vapour so the density altitude obtained above, or from the use of a flight computer, is for dry air. Water vapour in the atmosphere makes air less dense and for some critical operations it must be considered. This is done using the concept of virtual temperature.

37. The "Virtual Temperature" is the temperature that dry air would have so that its density would be the same as the moist air being considered. Since moist air is less dense than dry air, the virtual temperature will always be higher than the temperature of the air.

38. The amount of moisture in the air can be assessed by its dew point. The following table can be used to obtain the virtual temperature.

| Dew Point | Add to air temperature to obtain virtual temperature. |
|---|---|
| less than 10°C | 1 |
| 11°C to 20°C | 2 |
| 21°C to 25°C | 3 |
| over 2°C | 4 |

39. If the dew point in the example in paragraph 35 were 12°C, the virtual temperature would be 22°C and the density altitude would be 7,040 feet, (120 x 17) + 5,000.

# Summary - Chapter 13

- Pressure Altitude is the reading of the altimeter when 29.92 inches is set on it. It is the altitude designated for each pressure in the International Standard Atmosphere (ISA).

- Density Altitude is that altitude in the ISA corresponding to the prevailing air pressure and ambient temperature. It is the pressure altitude corrected for temperature.

- Indicated Altitude is the reading of the altimeter when the local altimeter setting is placed on the subscale.

- True Altitude is the indicated altitude corrected for flight level temperature. It is correct only if the lapse rate is 2°C/1,000 feet.

- Altimeter Setting is used in low-level airspace and when set on the subscale will cause the altimeter to read station height when the aircraft is on the aerodrome.

- Standard Pressure Setting is 29.92 inches of mercury set on the subscale. The altimeter will read pressure altitude and is used to indicate flight levels in the high-level airspace. It is also used in the standard pressure region.

- Surface Pressure Error - "From high to low look out below." It is particularly pertinent in the low-level standard pressure region using a fixed altimeter setting of 29.92 inches. A decrease of 1 hectopascal of surface pressure lowers the altitude by about 30 feet. This error is the same at all levels, and is corrected by using the altimeter setting.

- Temperature Error - The altimeter overreads in cold air and underreads in warm air. A correction can be made using OAT and a flight computer or using four feet per degree from ISA per thousand feet. IT IS ONLY CORRECT IF ISA LAPSE RATE OCCURS BELOW THE AIRCRAFT. This error increases with altitude and reduces to zero at ground level at the point where the pressure was measured.

- Combined Error - Cold temperatures plus low surface pressures combine to make excessively large total errors. These errors can be particularly significant in the low-level standard pressure region.

**13** CHAPTER

# Chapter
# 14

Strange things happen in the airflow near ridges and mountains. Although you are holding altitude and power, you may find your airspeed, vertical speed and altitude fluctuating widely and you may suddenly enter violent turbulence. You are in mountain waves.

# CHAPTER 14

# MOUNTAIN WAVES

## Introduction

1. The airflow over a mountain or a hill is disturbed in a similar way to that of a flow of water over a rock submerged in a river. The water rises over the rock then dips sharply on the downstream side, then rises and falls in a series of waves downstream. The crests of the ripples form a series of bars parallel to the rock that remain stationary with the water flowing through them. The ripples on the river surface are analogous to wave-like oscillations of the atmosphere which sometimes occur over and in the lee of mountains or hills. These waves are called "Mountain Waves" or "Lee Waves."

2. The extent of the oscillations depend on conditions in the atmosphere and on the size and shape of the topographical features causing them. The greater the height of the hill or mountain and the sharper the drop-off on the leeward side, the more extensive the oscillations. A ridge many miles long has a much greater effect than an isolated hill or peak. The location of other ridges downwind may amplify or nullify the wave motion depending on whether they are under a crest or a trough. Mountain ranges develop the strongest waves but they can also form with hills 300 feet high.

3. The stronger the wind, the longer the wave length. They vary from around 3 miles to 15 miles with 6 miles being a common average. The amplitude of the wave can be as much as 3,000 feet with vertical currents occurring up to 5,000 feet per minute. The wave amplitude normally increases from the surface to mid-tropospheric levels then decreases again except for large mountains and strong winds when the maximum amplitude is in the upper troposphere. The wave next to the ridge is normally the strongest and they get progressively weaker downstream from the ridge. They may extend for 300 miles in extreme cases.

4. The waves are predominant in the troposphere but they may reach into the stratosphere and have even occurred up to 70,000 feet. The wave length is longer at these higher levels than in the troposphere.

5. A large eddy called a "Rotor" may form under the wave crests between the friction layer along the earth's surface and the free flowing air in the wave flow (Figure 14-1(c)).

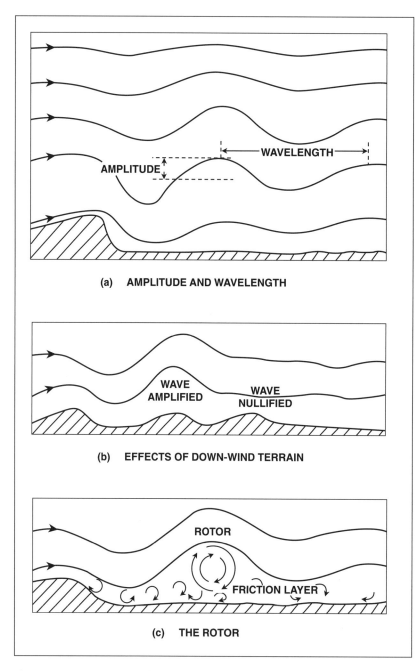

(a)  **AMPLITUDE AND WAVELENGTH**

(b)  **EFFECTS OF DOWN-WIND TERRAIN**

(c)  **THE ROTOR**

**Figure 14-1 Features of mountain waves**

## Clouds

6.  The atmosphere frequently has stratified moist layers in it. Very distinctive clouds will form in these layers in a lee wave flow.

7.  Orographic lift may form a cap cloud as air is forced up the windward face of the mountain. The cloud base is usually near the peak of the ridge with the cloud top a few thousand feet higher. The wind carries the cloud down the lee slope where the cloud dissipates by adiabatic heating. The cloud in this area assumes a waterfall appearance which indicates that the wind is rushing down the lee slope.

8.  Wave crests may be marked by lenticular (lens-shaped) clouds. Since the waves are stationary with the air flowing through them, the cloud forms on the upwind side and dissipates on the downwind side. These clouds form frequently in lee wave conditions and are one of the most common clues that lee waves are present. They form in the moist layers and do not necessarily indicate the level of maximum amplitude of the waves. They are most frequent at mid-cloud levels where they are called altocumulus standing lenticulars (ACSL), but can also form at high cloud levels. High cloud is composed of ice crystals which do not dissipate as easily as water droplets so the cloud may not dissipate on the downwind side of the wave. In this case, the cloud will simply have a very wavy appearance and will not be lens-shaped.

9.  A rotor cloud may form in association with the rotor. It will appear as a long line of stratocumulus parallel to the ridge and lying stationary a few miles from the ridge. Its base will normally be below the peaks, but its top can extend above them.

10. Even if lee waves are present, these distinctive clouds will not form if the air is dry or they may be embedded in other clouds and not be recognizable if the air is very moist. Generally, however, they are an excellent clue that waves are present.

Photo courtesy of NASA Langley Research Center

**Figure 14-2 Lenticular clouds (ACSL)**

## The Formation of Mountain Waves

11. Lee waves will form if the wind direction is within 30° of the perpendicular to the ridge. The wind speed at the summit must exceed 15 knots for small ridges and 30 knots for higher mountain ridges. The wind speed should increase and the direction remain constant with height up to the tropopause. The air should be stable near the mountain top level, but less stable above and below this. A jet stream nearly perpendicular to a mountain range provides ideal conditions for pronounced mountain waves. In this situation the wind increases with height up to the jet and its underlying associated frontal surface will provide a stable layer.

## Mountain Wave Turbulence

12. The turbulence associated with the rotor is extremely hazardous and can be as severe as any turbulence encountered in the atmosphere. The turbulence within the cloud is the worst, but it does extend outside the cloud. The strongest rotor normally occurs with the wave nearest the ridge.

13. How this turbulence develops is illustrated in Figure 14-3. If the rotor forms within an inversion, warm air from the top of the inversion will be rotated downward and heated further by compression while cold air from the bottom of the inversion will be carried upward and be cooled further by expansion. This results in very cold air lying over much warmer air. The condition now has become extremely unstable and it literally erupts in turbulence violent enough to cause structural damage to any aircraft.

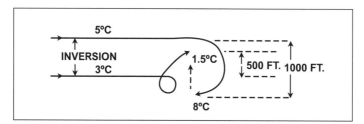

**Figure 14-3 Development of rotor turbulence**

14. Figure 14-4 shows the airflow through a lee wave. Because of the requirement for stable air the flow through the waves tends to be laminar. One of the features of lee waves is the noticeably smooth flight outside of turbulent areas. As shown in the illustration, the wind is stronger through the wave crest than elsewhere. If the shear becomes too great, the smooth flow will break into turbulence. This flow is evident in the lenticular clouds. Where the flow is laminar, the clouds are silky smooth in appearance; where it is turbulent, the clouds appear ragged and wind-torn. The transition from smooth to turbulent flight can be abrupt and moderate to severe turbulence can occur.

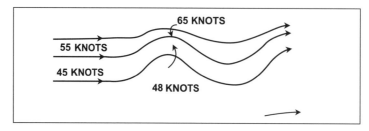

**Figure 14-4 Airflow through a lee wave**

15. Areas of strong jet stream shear that are overlying or embedded in a lee wave flow are more prone to turbulence than elsewhere along the jet stream.

16. The flow in the friction layer on the upwind slope is normally smooth. However, on the lee slope there can be severe turbulence as was explained in Chapter 11. A waterfall appearance on the lee side of the cap cloud is evidence of strong winds blowing down the slope and causing severe mechanical turbulence.

17. Very severe turbulence can also occur under the rotor. Smoke and dust lifted by the wind and evidence of a reversal of airflow under the rotor imply that turbulence is occurring.

## Temperature

18. The vertical motion in lee waves causes the air to undergo adiabatic expansion and compression with the result that the temperature varies markedly. The temperature at a particular level will be cooler in the crests than in the troughs. The freezing level, for instance, will be lower in the crests and higher in the troughs and in powerful waves the difference can be as much as 3,000 feet.

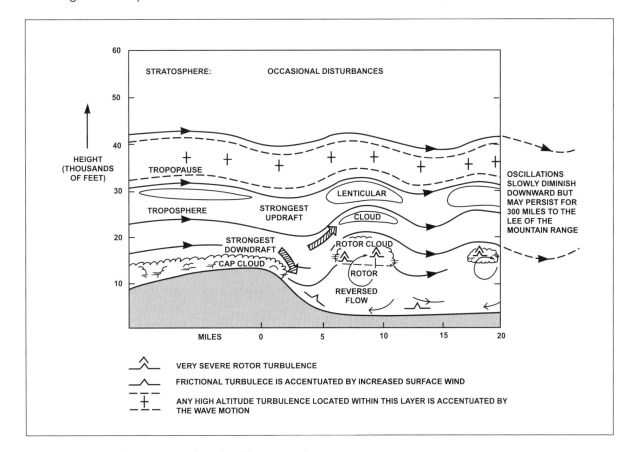

**Figure 14-5 Conditions associated with mountain waves**

**Effects on the Aircraft**

19. The structure of mountain waves causes a variety of effects on aircraft. The airspeed will fluctuate as the aircraft passes from the high wind speeds in wave crests to the slower speed in troughs. If a head wind abruptly increases, the airspeed will increase, if the head wind decreases; the airspeed will decrease. In a tailwind situation, an abruptly decreased tail wind will cause the airspeed to increase and an abruptly increased tail wind will cause the airspeed to decrease.

20. The vertical currents will also affect airspeed if a constant altitude is flown without changing power. An upcurrent will cause an increased airspeed, a downcurrent a decreased airspeed.

21. Mountain wave turbulence has been described and its effect can vary from being only a nuisance to catastrophic. The rotor is the most turbulent area.

22. Because of the large vertical currents, the liquid water content of clouds in lee waves is very high with the result that icing can be very serious.

23. The combined effects of airspeed fluctuations, turbulence and icing increase the chances of either stalling or overspeeding the aircraft.

24. The fact that the waves are stationary means that their effect on an aircraft is different when flying into wind than when flying downwind. The speed of the aircraft through the waves is greater flying downwind so it will not be affected as long by the vertical currents. It will, however, hit the turbulent areas faster and therefore receive more violent turbulence. When flying into wind, the speed through the turbulent areas will be less and so the turbulence will be less severe. The aircraft, however, will be in the vertical currents longer.

25. The increased wind speed in the wave crests causes a local lowering of the pressure (Bernouilli's Principle). This decreased pressure causes an error in the altimeter reading such that it indicates a height greater than you are actually at. This error combined with other altimeter errors can be as great as 3,000 feet.

26. Downdrafts are most severe near a mountain and at about the same height as the mountain top. An aircraft flying parallel to a mountain ridge on the downwind side may enter a smooth but strong down-draft. Due to the local drop in pressure associated with the wave, both the rate of climb indicator and the altimeter may not indicate a descent until the aircraft actually descends through a layer equal to the altimeter error. In fact, both may indicate a climb for part of the descent, and it is possible for an aircraft to be forced onto the ground in a situation such as this.

27. If possible, fly around the area when wave conditions exist. If this is not feasible, fly at a level which is at least 50% higher than the height of the mountain range. Be cautious to attain or maintain the minimum safe altitude during climb-out to cruising altitude, and descents for landings. This procedure will not keep you out of turbulence, but will keep you away from the mountain.

**CHAPTER 14**

28. Avoid the rotor clouds since they are the areas with the most intense turbulence of the mountain wave.

29. Avoid the strong downdrafts on the lee side of mountains. Approach the ridge at a 45° angle so that a quick turn away from the mountain can be made if necessary.

30. Penetrate turbulent areas at airspeeds recommended for your aircraft.

## Recognition of Mountain Waves

### Pre-Flight

31. If you are about to fly through an area where the terrain may cause lee waves, you should check the aviation forecasts to see if lee waves or lee wave turbulence is forecast. As with CAT, general areas of wave occurrence can be forecast, but because of extremely complicated terrain features, specific locations may not be. Weather reports should also be checked, particularly to see if altocumulus standing lenticulars (ACSL) is being reported. PIREPs of waves and turbulence are an excellent source of information. Wave clouds frequently stand out prominently on satellite photographs so if these are available, they should also be checked. The upper-air flow should be checked on the upper-air charts to see if the flow is cutting across ridges nearly at right angles. You should be cautious if winds are in excess of 50 knots and particularly if a jet stream cuts across the area.

### In-Flight

32. Exceptionally smooth flying conditions outside of turbulent areas are a noticeable feature of lee wave situations. Fluctuations of the airspeed indicator and vertical speed indicator may occur. The presence of cap, lenticular and roll clouds imply lee wave conditions. Surface smoke and dust rising up several thousand feet can be related to a rotor when a roll cloud is absent.

**14** CHAPTER

## Summary - Chapter 14

- Conditions most conducive to lee wave formation are: a long high ridge with an abrupt drop off on the lee side, strong winds nearly perpendicular to the ridge increasing in speed with height but not changing direction, and a stable layer near the ridgetop height.

- The wave nearest the ridge is the strongest. Waves can extend up to 300 miles downwind and occasionally reach into the stratosphere.

- Rotor, cap and lenticular clouds are evidence of lee waves.

- In-flight evidence of lee waves include: exceptionally smooth flying conditions, fluctuations in airspeed, vertical speed and altimeter indicators while holding a constant attitude, the presence of lee wave clouds, and smoke and dust rising up from the surface.

- Pre-flight evidence of lee waves include forecasts, weather reports of ACSL, PIREPS, satellite data, and the indication of the upper-air flow from upper-air charts.

- Hazards include turbulence, especially in the rotor, icing, loss of control of the aircraft due to a stall or overspeeding, and errors in the altimeter and vertical speed indicators.

- Flight procedures include flying at least 50% higher than the mountain range, avoid the rotor cloud and the strong downdrafts on the lee side of the mountain, don't rely on altimeter readings and penetrate turbulent areas at appropriate airspeeds.

CHAPTER **14**

# Chapter
# 15

Thunderstorms pack a wallop that contains most of the severe atmospheric hazards. Since it is almost inevitable that you will encounter thunderstorms you should learn what these hazards are and how they should be handled.

# CHAPTER 15

## THUNDERSTORMS

### Introduction

1. A thunderstorm begins with the formation of a convective cloud type such as cumulus or altocumulus castellanus. If the air mass is conditionally or potentially unstable through a very deep layer, preferably extending up to the tropopause, and if it is very moist in its lower levels, the convective clouds may grow and merge into cumulonimbus (CB), the thunderstorm cloud. These clouds move with the general airflow within the troposphere, but set up small and violent circulations around and within themselves.

Photo courtesy of NASA Langley Research Center

**Figure 15-1 Cumulonimbus (CB)**

2. Drafts and gusts are prominent features of thunderstorms. Drafts can be either updrafts or downdrafts and are large enough that they can move the aircraft vertically. The aircraft can be wholly contained within the draft and although a thousand feet or more may be gained or lost there may be little turbulence.

3. Superimposed upon the large-scale continuous flow of the drafts are numerous irregular random, sudden and brief turbulent motions called "Gusts." They have a significant effect upon the aircraft, causing pitch, roll and yaw movements. The gusts vary in size from a few inches to whirling masses of air several hundred feet in diameter. They are produced by the shearing action between drafts and the air surrounding the drafts, or between adjacent updrafts and downdrafts. In the aircraft they can be felt as vertical or horizontal jolts.

15 CHAPTER

4.  The upward motion of the initial updraft in the cloud is accelerated by the release of latent heat, first as water vapour condenses to water droplets and then, higher in the cloud as water droplets freeze to become ice crystals. All thunderstorms progress through a life cycle from their initial development through maturity into a dissipating stage. This life cycle takes from one to three hours.

## The Cumulus Stage

### Updraft

5.  The first stage of a thunderstorm cell is the "Cumulus stage" (Figure 15-2(a)). Several small convective clouds will merge and begin to grow rapidly into a very large convective cell. The base may expand to four miles across and the cloud top may rise to over 20,000 feet. Only updrafts are present in the cloud and air is drawn into the sides as well as into the base of the cloud. The updraft speed is greatest in the upper levels and can reach 3,000 feet per minute.

### Rain

6.  As the cloud builds past the freezing level, ice crystals form and these cause raindrops to form. Because of the strong updraft the drops become very large and are carried to great heights. As the raindrops grow larger and heavier, they fall. Up to this point, the entire cell has been engulfed in an updraft. The average life of this stage is about 20 minutes.

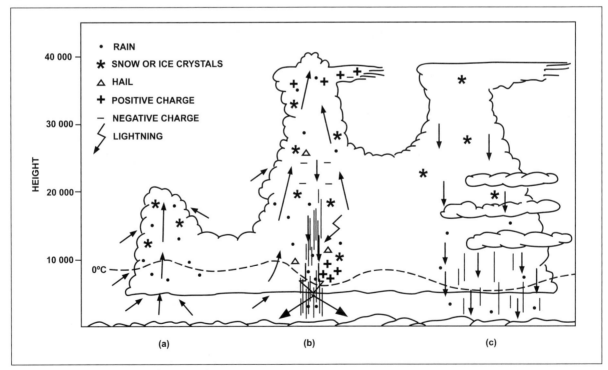

**Figure 15-2 The stages of a thunderstorm**

CHAPTER **15**

# The Mature Stage

## Updraft

7.  The beginning of rain at the earth's surface marks the transition of the cell to the mature stage (Figure 15-2(b)). The cloud continues to grow and updrafts reach their maximum speed, possibly reaching 6,000 feet per minute. The cloud reaches the tropopause and if the updraft is particularly strong, it may burst into the stable air of the stratosphere for 5,000 or 10,000 feet. The strong winds at the tropopause carry the cloud ahead of the storm, forming an anvil shape with the inversion above the tropopause acting as a lid to hold the anvil beneath it (Figure 15-3).

**Figure 15-3 Thunderstorm anvils**

## Downdraft

8.  As the raindrops fall, they drag air down with them. This, in conjunction with cooling of the air due to evaporation of the raindrops, starts a downdraft in the middle portion of the cloud that accelerates downwards. The speed of the downdraft may be as high as 2,500 feet per minute.

## Gust Front

9.  The cold air pours out of the cloud base, strikes the ground and spreads out radially. Because it carries the motion of the stronger upper winds with it, it spreads out ahead of the storm farther than elsewhere, reaching out to possibly 10 or 15 miles in front of the main cloud mass. This horizontal outflow of air produces strong and gusty surface winds accompanied by a sharp drop in temperature and an abrupt rise in pressure. The onset of this wind is a "Gust-Front."

## Downburst and Microburst

10. Ahead of the gust front, air is flowing into the storm by rising up and over the cold downflowing air (Figure 15-4). With the passage of the gust front, the wind will swing around, possibly 180 degrees and increase sharply as a squall. The downdraft may be so severe that damaging winds result, often called a "Downburst." On occasion, embedded within the downburst area are exceptionally violent winds that last for only a few minutes and that are only about a mile across. These are called "Microbursts." These winds have all originated as downdrafts that flow out and away from the storm.

## Roll Cloud

11. A "Roll Cloud" may form near the main cloud base in the shear area at the front of the cloud where the downdraft comes out of the cloud base and the updraft enters it.

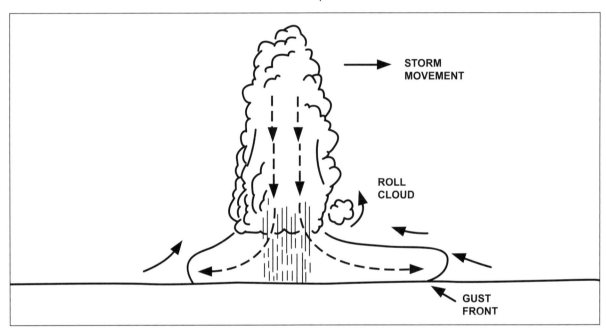

**Figure 15-4 The gust front and roll cloud**

## Hail

12. Hail is most likely to occur in the mature stage. Super-cooled raindrops above the freezing level begin to freeze. As they fall, other drops coalesce on them and freeze and the hailstone grows in size. If hailstones should fall into a strong updraft area, they are carried back aloft, continuously growing until they are too large for the updrafts to carry and they fall. As they fall below the freezing level, they begin to melt and may reach the ground as either rain or hail. For this reason, hail is most prevalent and larger above the freezing level. It is possible for smaller hail to be carried up in an updraft and thrown out of the side of a storm into the clear air several miles downwind.

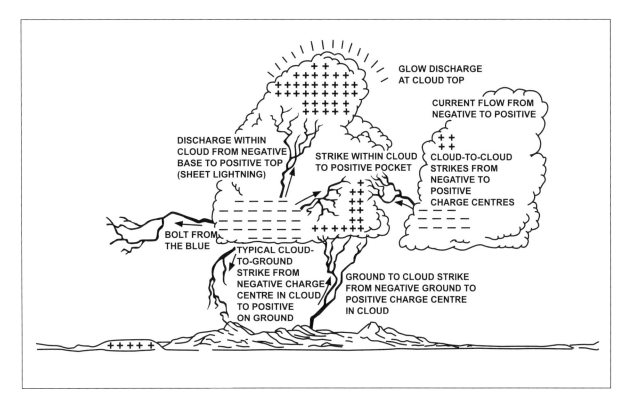

**Figure 15-5 Lightning**

## Lightning

13. The cloud is now composed of strong up and downdrafts occurring side by side. Water droplets, hail and snow are moving up and down and it is in these complex and changing conditions that lightning occurs. Areas of positive and negative charge accumulate in different parts of the cloud until the difference in electrical potential reaches a critical value. When this occurs, a lightning discharge takes place, sometimes from one part of the cloud to another, sometimes from one cloud to another, or from cloud to ground. Occasionally a discharge takes place from the top of the cloud to the atmosphere above the cloud and, on rare occasions, from the ground to the cloud.

14. Thunder is the noise made as a result of the lightning heating the air so rapidly that it expands faster than the speed of sound. The rumbling is due to the distance between the listener and the various parts of the lightning channel. The distance in miles to the lightning strike can be estimated by counting seconds between lightning and thunder and dividing that number by five.

15. The mature stage is the most violent stage of the thunderstorm cell and usually lasts for 20 to 30 minutes. It ends as the updraft weakens and the downdraft grows in size and encompasses the entire cloud.

**15** CHAPTER

## Severe Storm Structure

16. The more severe thunderstorms have a slightly different structure than has just been described. If horizontal wind speeds increase markedly with height, considerable tilt develops in the updraft and cloud. In this situation, precipitation falls through only a small portion of the rising air (Figure 15-6). Since the drag of falling raindrops is not imposed upon the rising currents in the cloud, the updrafts can continue. If there is a layer of dry air at mid levels through which the rain is falling, evaporative cooling enhances the downdraft speed. In the presence of this cold downdraft, the tilted updraft becomes part of an overturning of the air in a deep layer of the troposphere and the ascending air can reach velocities in excess of 10,000 feet per minute. It may penetrate several thousand feet into the stratosphere, then settle back into the anvil. Storms developing in this way become persistent and self-propagating and may develop tornadoes.

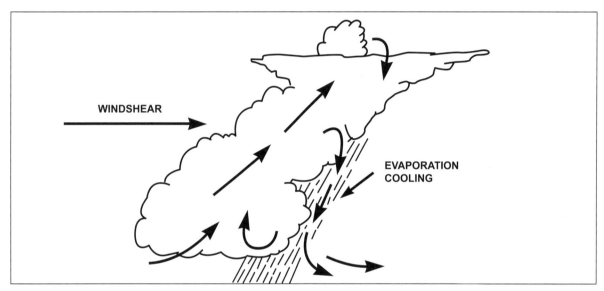

**Figure 15-6 The effect of wind shear on a thunderstorm**

## Tornadoes

17. Tornadoes are violent rotating columns of air that descend out of a thunderstorm in the shape of a funnel. A tornado vortex is several hundred yards in diameter with the wind rotating rapidly around it with extremely low pressure in the centre. If the vortex touches the ground, it is called a "Tornado," if it remains aloft hanging from the cloud base it is called a "Funnel Cloud." Either can extend upward into the cloud for over 30,000 feet. They are most common in the south and southwest parts of a thunderstorm and may enter the cloud in a line of innocent looking cumulus in a rain-free area extending several miles away from the parent storm (Figure 15-7). They tend to occur as families so if one is seen others are probably occurring. Over water a tornado forms a "Waterspout."

18. Tornadoes are not nearly as frequent in Canada as they are in the United States, but they do occur, particularly near the Canada - United States border, frequently enough to constitute a significant hazard.

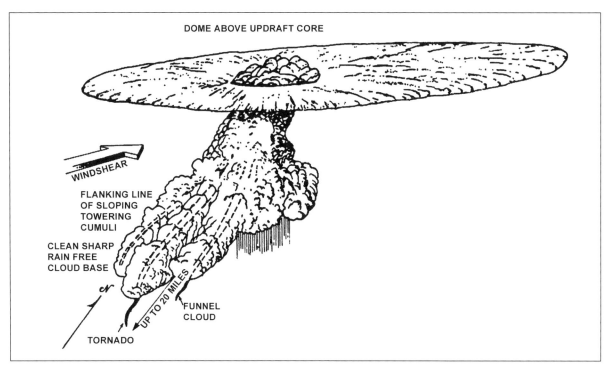

**Figure 15-7 Tornadoes**

## The Dissipating Stage

### Downdraft

19. Towards the end of the mature stage, the downdraft grows in size while the updraft weakens until the entire cell becomes an area of downdraft. The mechanisms which generated lightning, rain, cloud, hail and turbulence have been removed and the lower portions of the cloud frequently become stratiform in appearance and then dissipate. The anvil, however, may remain for a considerable time. The dissipating stage may last for one or two hours and is the most prolonged of the three stages of the thunderstorm cell and is also the least hazardous (Figure 15-2(c)).

20. An individual thunderstorm usually consists of two or more cells in different stages of their life cycles. It may cover an area from around 10 miles in diameter to as much as 100 miles (Figure 15-8).

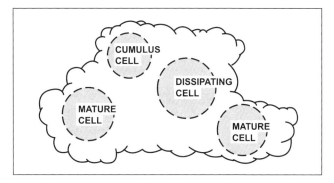

**Figure 15-8 Looking down on a thunderstorm**

# Classification of Thunderstorms

21. Thunderstorms are classified according to the trigger action that sets off the instability. The lifting action necessary for their formation may be furnished by any type of front, mountainous terrain, convective currents or convergence. The latter three sources of lift can be found either within an air mass far removed from a front or in conjunction with a front. Thunderstorms which are far removed from fronts, or troughs, tend to be widely scattered or isolated; those associated with fronts or troughs tend to be more numerous and form in lines. It is convenient, therefore, to identify them as either air mass or frontal thunderstorms. While it is not an infallible rule, you can expect more numerous thunderstorms when they are called frontal than when they are called air mass. The following paragraphs give a general description of the various thunderstorm types. In a particular case, it will be the characteristics of the atmosphere at that time that will determine the various storm features.

## Frontal Thunderstorms

22. Because of the shallowness of the frontal slope, stratiform clouds are likely to accompany warm fronts and thunderstorms may be embedded, and not be visible unless the pilot is flying above the stratiform layer. Because of the shallow slope of the warm front, they are usually the least severe of all frontal thunderstorms (Figure 15-9).

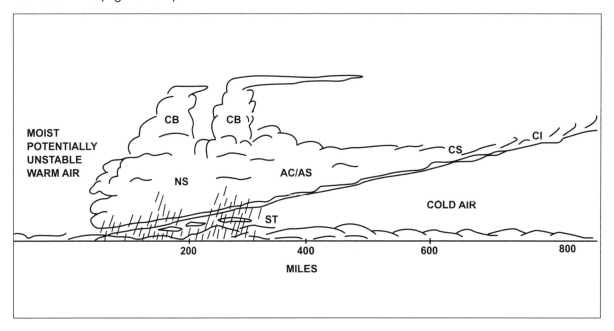

**Figure 15-9 Warm front thunderstorms**

23. Thunderstorms associated with cold fronts are normally the most severe ones found anywhere except in squall lines. They usually form in a continuous line and are easy to recognize by a pilot approaching the front from any direction. Their bases are normally lower than those of other frontal thunderstorms and they are usually most active during the afternoon (Figure 15-10).

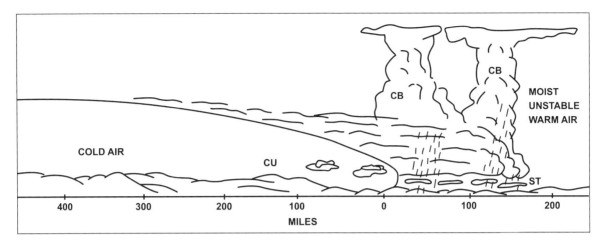

**Figure 15-10 Cold front thunderstorms**

24. Thunderstorms are often associated with a trowal. In this case, they occur along the upper cold front and are set off by the rapid lifting of warm, moist air. They may be more severe than warm front thunderstorms but, in similar fashion, are usually embedded in stratiform clouds (Figure 15-11).

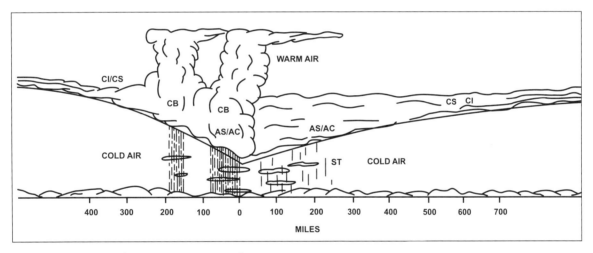

**Figure 15-11 Thunderstorm at a trowal**

## Squall Line Thunderstorms

25. Thunderstorms along a squall line frequently are similar to those along a cold front but may be more violent. The cloud bases are often lower and the tops higher than with most other thunderstorms. The most severe conditions, such as heavy hail, destructive winds and tornadoes, are generally associated with squall line thunderstorms. Usually most intense during the late afternoon and evening, squall line thunderstorms may occur at any time (Figure 15-12).

26. Chapter 8 indicated that squall lines often develop about 100 to 200 miles ahead of and roughly parallel to fast-moving cold fronts. Squall lines usually form rapidly, and sometimes a series of them will develop ahead of the cold front with a new squall line springing up to take the place of one which might move out rapidly in advance and dissipate.

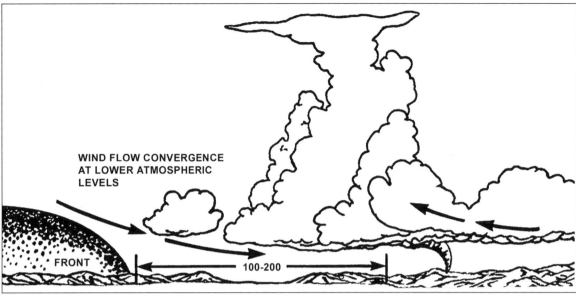

**Figure 15-12 Squall line thunderstorms**

27. While squall lines frequently accompany cold fronts, the existence of a front is not a prerequisite. They may accompany low-pressure troughs, or lines where sea breezes converge against mountain barriers. Other factors being favourable, squall line thunderstorms are most likely to develop in areas where there is a convergence of wind flow in the lower atmospheric levels, regardless of the cause of the converging flow.

## Air Mass Thunderstorms

28. The basic characteristics of air mass thunderstorms are that they form within a warm, moist air mass, and are in no way associated with fronts and that they are generally isolated or widely scattered over a large area. Occasionally, however, they do combine to form complexes of thunderstorms thousands of square miles in extent. Air mass thunderstorms may be classified as "Convective," "Orographic" or "Nocturnal."

## Convective Thunderstorms

29. Convective thunderstorms may receive their necessary lift by heating from below or by convergence of the wind flow. An example of convergence in the wind flow within an air mass is an occluded low pressure containing no fronts (Figure 15-13).

30. Those convective thunderstorms formed by convergence have no particular diurnal variation with respect to time of occurrence. However, they tend to be most active over land in the afternoon and early evening, and most active over water during the night and early morning. This diurnal variation in activity over land results from solar heating, and over water, by radiation cooling of the cloud tops.

31. Thunderstorms receiving their lift only through heating from below seldom are found in an appreciable wind flow. Strong winds tend to break up the convective currents.

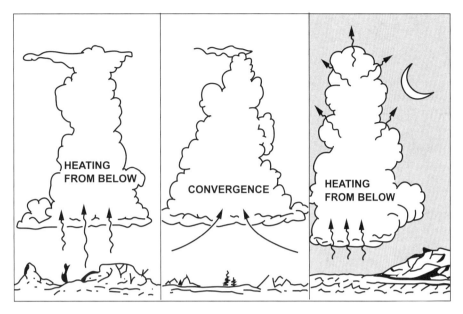

**Figure 15-13 Convective thunderstorms**

## Orographic Thunderstorms

32. Orographic thunderstorms develop when the wind forces moist, unstable air up mountain slopes. They tend to be more frequent during the afternoon and early evening because heating from below is working in conjunction with the forced lifting. The storm activity is usually widely scattered along the individual peaks of the mountains, but occasionally there will be a long unbroken line of thunderstorms. Violent thunderstorms with hail are common in high mountains such as the Rockies (Figure 15-14).

**Figure 15-14 Orographic thunderstorms**

**15** CHAPTER

33. Identification of orographic thunderstorms from the windward side of the mountain is often difficult during low-level flight. Stratus or stratocumulus clouds frequently enshroud the mountains and obscure the storm clouds.

## Nocturnal Thunderstorms

34. The term nocturnal is applied to a peculiar type of air mass thunderstorm found in the Midwest. It frequently occurs at night or early in the morning in the Central Plains area. These thunderstorms are associated with unusually moist air aloft. The trigger action initiating these storms is thought to be night-time radiation from this moist air layer.

# Thunderstorm Hazards

35. Thunderstorms may be accompanied by severe or extreme turbulence, icing, lightning, thunder, precipitation and gusty surface winds. The more severe ones produce hail and sometimes tornadoes.

## Turbulence

36. Turbulence is hazardous because of the danger of losing control of the aircraft and of the aircraft being overstressed. Most thunderstorms possess the potential to produce extreme turbulence.

37. The greatest turbulence is likely to be encountered when traversing adjacent rising and descending drafts. Gustiness is usually greatest in these shear zones between opposite moving drafts. On weather radar, there is a correlation between turbulence and the rainfall gradient. There will be turbulence associated with areas where the rainfall changes from no rain or light rain to heavy rain in a very short distance.

38. Updrafts are generally stronger and larger in both horizontal and vertical extent than downdrafts. They are strongest in the middle and upper levels of the thunderstorm where upward displacement of aircraft of up to 6,000 feet has been encountered, but more often this displacement is less than 3,000 feet.

39. Strongest downdrafts aloft are found in the middle levels of the thunderstorms. While updrafts are generally stronger than downdrafts, downward displacements of aircraft as much as 8,000 feet have been encountered.

40. Downdrafts present the additional hazard of sometimes being present to ground level in the first gust, as we shall see later. For this reason great caution should be exercised if attempting to fly beneath the thunderstorm clouds, especially over irregular terrain.

## Turbulence and Altitude

41. Variation of turbulence intensity appears to be almost constant throughout the storm. However, there is a slight tendency to show a maximum intensity near the mid-level of the storm.

## Turbulence and Distance from the Storm Cell

42.  The storm cloud is only the visible portion of a turbulent system that extends outside the cloud itself. The severity of turbulence associated with severe storms decreases slowly with distance from the storm to about 20 miles from its centre although this distance will be greater downwind than into wind. Severe turbulence can be encountered in the anvil up to 30 miles downwind. The recommended technique using radar is to avoid thunderstorm returns by 5 miles when flying below the freezing level, 10 miles when flying above the freezing level and 20 miles when flying above 30,000 feet. No flight path, through an area of strong or very strong radar echoes separated by 40 NM or less, can be considered free of severe turbulence.

## Turbulence Above Storms

43.  An aircraft at the very high levels required to top severe storms may be flying close to its stall speed. A stall at this altitude may take several thousand feet to recover and the aircraft may enter the thunderstorm. This could have serious consequences. Turbulence does occur above storms and if it is encountered, a stall could result. As a general rule you should clear the top of a storm by 1,000 feet for every 10 knots of wind speed at cloudtop level.

## Turbulence Below Cloud Bases

44.  Although turbulence may appear to be at a maximum at mid-levels, as discussed earlier, it can also represent a hazard at lower altitudes. Flying at 4,000 to 6,000 feet AGL under the cloud base can lead to very severe turbulence encounters in severe thunderstorm situations.

## Turbulence and Visual Appearance of a Storm

45.  There is no useful correlation existing between the external visual appearance of a thunderstorm and the turbulence and hail within it. There is a relationship, however, between turbulence and the amount of lightning. Normally, the more lightning a storm produces, the more turbulent it is.

## Hail

46.  Hail competes with turbulence for first place as the greatest thunderstorm hazard to aircraft. It can cause severe airframe damage particularly along leading edges, radomes and windshields. Most, and perhaps all, thunderstorms have hail in the interior of the cumulonimbus cloud at some stage in their lives. In a large percentage of the cases, the hail melts before reaching the ground, but this does not lessen its danger to the pilot who encounters it aloft.

47.  Large hail is most commonly found in severe thunderstorms. It is usually produced during the mature stage of cellular development and is most frequently encountered at levels between 10,000 and 30,000 feet. Smaller hail can be carried up in an updraft and thrown out of the side of the storm into the clear air ahead of it

**15**

CHAPTER

## Rain

48. The thunderstorm contains considerable quantities of liquid water, but this moisture is not necessarily falling to the earth as rain. Raindrops are suspended or carried aloft by the updrafts. Rain may be encountered up to 40,000 feet even when the outside air temperature is well below freezing. Thunderstorm penetration at high speeds indicates that this water presents as much danger to the aircraft as hail, mainly because water encounters are more likely than hail. Impact pressure created by water has been computed to be 18,000 pounds per square inch when flying at a speed of 1.6 Mach number. This pressure has been known to peel flush rivet heads out of the aircraft's leading edges; plexiglass is worn down, fibreglass antennas eroded away, and paint peeled off. To keep water damage to a minimum, the aircraft must penetrate the thunderstorm at the recommended airspeed, not faster.

49. Heavy rain showers aloft can attenuate airborne radar so that it is unable to see through the severe parts of a thunderstorm to what is lying behind. Heavy showers can also cause loss of power and possible flame-out of jet engines, particularly if they are throttled back to idle power during descent.

50. During approach and landing, heavy rain can reduce visibility and cause refraction as it impinges on the wind screen. This gives an illusion that the runway threshold is lower than it actually is. There is also evidence that heavy rain showers may cause loss of lift due to roughening of the upper wing surface, which increases the stall speed. Water lying on the runway can cause hydroplaning with loss of control of the aircraft during its landing roll.

## Icing

51. Clear icing in cumulus clouds and thunderstorms can accumulate rapidly. The heaviest icing conditions usually occur just above the freezing level where the greatest concentration of super-cooled water drops exists, but severe icing may occur at any point above the freezing level at temperatures from 0°C to -25°C.

52. In areas where individual thunderstorms are widely dispersed, icing usually does not present too serious a problem because the flight time in each storm is relatively short. In areas where thunderstorms are clustered, the icing problem may be serious if the aircraft has prolonged exposure to icing conditions.

## Altimetry

53. During the passage of a thunderstorm, rapid and marked surface pressure variations generally occur in a particular sequence characterized by: a) an abrupt fall in pressure as the storm approaches; b) an abrupt rise in pressure associated with rain showers as the storm moves overhead (very often associated with the first gust); c) a gradual return to normal pressure as the storm moves on and the rain ceases. Such pressure changes may result in significant altitude errors on landing. Because of the very local nature of the pressure variations, even a current altimeter setting as sensed at an air traffic control facility may not be accurate for a threshold some miles away. The error can be in the order of 100 to 150 feet.

# Lightning

54. Lightning strikes are of considerable significance to aviation. Statistics indicate that lightning strikes have been the cause of over 50% of all weather related mishaps that either caused accidents or had the potential to cause accidents. The probability of a lightning strike to an aircraft increases with increased aircraft size and increased speed. Although generally only minor damage occurs, there have been several aircraft destroyed by lightning. This destruction was the result of fuel vapour explosions or loss of flight controls or loss of flight instruments.

## Triggered Lightning

55. There are two types of lightning strikes that occur in flight. By far the most common variety is triggered by the aircraft itself. This can occur when a static electric charge builds up on an aircraft and it then flies through a strong electric field in the atmosphere. The aircraft can build up a static charge by flying through dry particles such as ice crystals or snow, or by flying through a mixture of rain and snow near the freezing level. Strong electric fields exist in the atmosphere nearby cumulonimbus clouds or clouds from which either rain or snow is falling.

56. As the static charge builds up on the aircraft prior to a strike, there will be a buildup of static noise in high frequency or low frequency communications equipment. At night, a corona may be seen across the windscreen or at extremities of the aircraft. This buildup will continue for several seconds, then a discharge (lightning) occurs between the aircraft and the cloud producing the electric field. Note particularly that a thunderstorm need not be present. If there is a cumulonimbus nearby, it may not have produced lightning prior to the aircraft strike.

57. Aircraft flying in the tops of cumulonimbus clouds where the temperature is less than -40°C appear to be particularly prone to aircraft-triggered lightning strikes.

## Natural Lightning

58. The second type of lightning strike occurs abruptly with no warning and is quite rare. It is a true lightning strike that occurs when an aircraft is near the path of a naturally occurring stroke of lightning from a thunderstorm. The voltages and current flow with a natural lightning strike is far greater than with a triggered strike. The lightning generally enters at one extremity of an aircraft such as the nose or wing tip, passes through the aircraft and leaves at another extremity, such as the tail. The stroke then continues on to the ground, or to another cloud, or to a different part of the same cloud. Frequently, a lightning flash is composed of several strokes with each stroke following the same path as the previous one. As an aircraft flies through this path, it may be hit by several strokes that leave a trail of small burn or pit marks along the fuselage as each new stroke strikes the aircraft. These are called "Swept Strokes."

**15** CHAPTER

# Lightning Damage

## Direct Effect

59. There are both direct and indirect effects of a lightning strike. The direct effects include physical damage due to arcing of the very large electric current associated with lightning. Since the lightning enters and leaves the aircraft at its extremities, these are the parts normally damaged. There is particular concern that composite type skins can be severely damaged because of their inability to carry the large current of a lightning strike.

## Indirect Effect

60. The indirect effects of a lightning strike can be as serious as the direct effects. There is a magnetic field associated with the lightning and this can induce currents in various aircraft components and avionics. This can occur even with nearby lightning that does not strike the aircraft. Damage to aircraft electrical systems, instruments, avionics and radar is possible. Lightning strikes have produced surges in the electrical system causing the generators to drop off the line. Lighting, magnetic compasses, and all electrical equipment could be lost. Induced currents could also actuate electrical circuits inadvertently so that, for example, external stores can be released. Aircraft incidents have indicated that lightning strikes have caused loss of the attitude indicator, the forward looking radar, the TACAN and the inertial navigation system (INS) all at the same time. Loss of the communications radio could further complicate the situation. Instrument failure or malfunctions should be expected because this failure may or may not be apparent to the pilot. All instruments should be considered invalid until proper operation is verified. Advanced guidance and control systems with their sensitive solid state electronics and micro-processors are highly susceptible to the induced currents and magnetic fields of lightning.

## Fuel Ignition

61. There have been occasional catastrophic fuel explosions. In general, JP-4 fuel is significantly more vulnerable to explosions at temperatures near 0°C where most strikes occur than either kerosene or gasoline.

62. Extensive research into this problem has resulted in improved fuel system designs that are lightning-resistant so that newer aircraft should be able to experience lightning strikes with no adverse effects on the fuel system.

## Effects on the Crew

63. There have been incidents where crews have felt firm jolts in their helmet area and their hands tingled and their hair stood on end. Incidents have been reported where the pilot received severe burns to the hands and face. There have been many instances of crews being temporarily blinded by lightning flashes at night. Flash blindness may last for 30 seconds or more. Pilot spatial disorientation, cockpit distractions and confusion are also possible and should be expected.

## Engines

64. Flame-outs or compressor stalls of fuselage-mounted engines have occurred. These have resulted when a hot lightning channel has swept in front of the engines disrupting the air flow into them.

## Effects on Radio and Navaids

65. Lightning discharges cause loud crashes of static, particularly on low, medium and high frequency radio receivers. Aircraft flying through snow, ice crystals or dust accumulate an electric charge that results in a steady level of noise called "Precipitation Static" on these same frequencies. In both cases there will be interference of radio reception or navigation using ADF. Very high frequency or ultra high frequency radios are not seriously affected. A brush or corona discharge may be visible at night on various parts of the aircraft while precipitation static is occurring.

66. Precipitation static near or within thunderstorms tends to be most severe in the region where the greatest weather hazards to aircraft are present; that is, near the freezing level where snow changes into rain or rain into snow, and severe turbulence along with updrafts and downdrafts are present.

## The Gust Front

67. A particularly significant thunderstorm hazard that has been responsible for some of the worst aircraft accidents is the "Gust Front." This hazard may affect take-off and landing near a thunderstorm up to an altitude of about 6,500 feet above ground level. The strong gust wind results from the horizontal spreading out of the thunderstorm's downdrafts as they approach the surface of the earth. The first gust usually is the strongest wind observed at the surface during the thunderstorm's passage. In downbursts, it may approach 100 knots. It has been found that the wind speed may increase as much as 50% between the surface and 1,500 feet with the greatest increase taking place in the first 150 feet. This is most significant because the surface observation may not give a true estimate of the actual wind speed just above the surface. At times, secondary or tertiary gust fronts are found between the first gust and the rain shaft due to microbursts in the thunderstorm.

68. Extreme wind direction change may also be encountered. The total first gust speed is the sum of the speed of the horizontally spreading downdrafts and the forward speed of the thunderstorm. Thus, wind speeds at the leading edge of the storm are considerably greater than those at the trailing edge.

69. As shown in Figure 15-15, the extremely turbulent first gusts precede the arrival of the thunderstorm, stirring up dust and debris, and making its approach easily visible. Close behind it, if present, is a long, rolling cloud just above the surface called the "Roll Cloud." Behind it is the main body of the thunderstorm. Strong gusty winds will last for a few minutes after the thunderstorm's passage because the downrush of air is spreading out in all directions. This wind condition, however, is not as dangerous as the first gust.

**15** CHAPTER

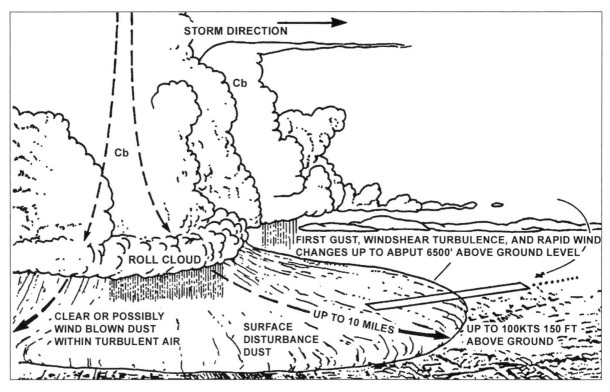

**Figure 15-15 Gust front hazards**

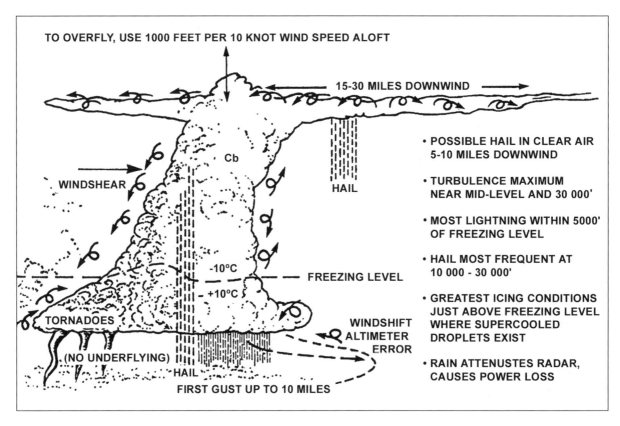

**Figure 15-16 Thunderstorm hazards**

## Flight Procedures

70. Avoid thunderstorms if possible. Try not to fly in layered cloud without radar if there is evidence of embedded thunderstorms from actual or forecast conditions. Use radar as a thunderstorm avoidance tool, not as a penetration tool. An early enroute detour is preferable to late detouring of individual storms. If you cannot avoid penetrating a thunderstorm, here are some dos before entering the storm:

    a. Tighten your safety belt, put on your shoulder harness if you have one, and secure all loose objects.

    b. Plan your course to take you through the storm in a minimum time and hold that course.

    c. Establish a penetration altitude below the freezing level or above the -25°C level to avoid the most critical icing.

    d. Turn on pitot heat and carburetor or jet inlet heat. Icing can be rapid at any altitude and can cause almost instantaneous power failure or loss of airspeed indication.

    e. Establish power settings for reduced turbulence penetration airspeed recommended in your aircraft manual. Reduced airspeed lessens the structural stresses on the aircraft.

    f. Turn up cockpit lights to highest intensity to lessen danger of temporary blindness from lightning.

    g. Disengage altitude hold mode and speed hold mode. The automatic altitude and speed controls will increase manoeuvres of the aircraft, thus increasing structural stresses in using automatic pilot.

    h. Tilt the radar antenna up and down occasionally. Tilting it up might detect a hail shaft that will reach a point on your course by the time you do. Tilting it down might detect a growing thunderstorm cell that might reach your altitude.

71. Dos and don'ts during thunderstorm penetration:

    a. Do keep your eyes on your instruments. Looking outside the cockpit can increase danger of temporary blindness from lightning.

    b. Don't change power settings. Maintain settings for reduced airspeed.

    c. Don't overstress the aircraft trying to maintain a constant altitude; let the aircraft ride the waves. If in radio contact, advise ATC of your inability to maintain a constant altitude.

    d. Don't turn back once you have committed yourself and are in the thunderstorm. A straight course through the storm most likely will get you out of the hazards quickest. In addition, turning manoeuvres increase stress on the aircraft.

**15** CHAPTER

e. Don't take off or land in face of an oncoming thunderstorm. You could encounter the downburst and shear area.

f. Do stay clear of very heavy rain shafts. They could cause loss of power and loss of lift.

72. Points regarding visual flight through thunderstorm areas:

a. The storm cloud is only the visible portion of a turbulent system that often extends outside of the storm proper.

b. If you are forced to climb to an altitude approaching the aircraft's ceiling to overfly the storm, there is great danger of engine flame-out or stalling if turbulence is encountered. Overfly by 1,000 feet for every 10 knots of wind speed at cloud top level.

c. The horizontal mass of cloud is greatest near mid and low levels and least at high levels so there is more clear air at high levels in which to circumnavigate.

d. Hail can be thrown out of the anvil into the clear air downwind of the storm.

e. Circumnavigate the storm by a minimum of 10 miles on the downwind side and a minimum of 3 miles on the upwind side.

f. If you are caught under a storm, be prepared for strong updrafts in rain-free areas and strong downdrafts in rain areas.

g. Watch for swirls of dust on the ground or hanging cloud elements in the base of the cloud: these could indicate tornado activity. The rain-free southwest portion of a storm is particularly prone to tornado development.

h. If one tornado is seen, expect others, since they tend to occur in groups.

i. Hail cannot be differentiated from rain visually.

j. The gust front may be identified by a line of dust and debris blowing along the earth's surface.

k. Don't attempt to fly under a thunderstorm even if you can see through to the other side. Turbulence under the storm could be disastrous.

l. Virga falling and evaporating from high-based storms can cause violent downdrafts.

# Summary - Chapter 15

- Thunderstorms are composed of several cells. They move with the general tropospheric flow but develop small violent circulations around themselves.

- The cells go through three stages:

    - The Cumulus Stage  – an updraft throughout.

    - The Mature Stage:  – rain starts,
        – downdraft develops, in severe cases they are called a downburst or a microburst,
        – hail can occur,
        – lightning occurs,
        – gust front develops from the downdraft,
        – tornadoes can occur,

    - The Dissipating Stage – a downdraft throughout.

- The storms are classified according to the trigger action forming them.

- Frontal thunderstorms - Cold front and squall line storms are generally the most severe.

- Air mass thunderstorms are caused by surface heating, orographic lift or cooling of a moist layer aloft at night.

- Turbulence can be extreme and can occur both in and outside the cloud.

- Hail occurs with most thunderstorms but frequently melts before reaching the ground.

- Rain can occur even at very high levels and can cause impact damage on the airframe, power loss or flame-out. It also causes refraction and hydroplaning on landing.

- Icing can be severe to -25°C.

- Lightning strikes can be triggered by the aircraft without a cumulonimbus being present, or can emanate from a cumulonimbus. Occasionally aircraft are struck by naturally occurring lightning.

- Large altimeter errors occur and are very local in nature so that an ATC altimeter setting may not be appropriate in touchdown area.

- Precipitation static causes HF, MF and LF radio interference.

- A wind shift, strong winds and turbulence are associated with the gust front.

- If a thunderstorm encounter is inevitable, there are flight procedures that should be followed that will reduce the risks.

**15** CHAPTER

# Chapter
# 16

Some of the most serious aircraft accidents in recent times have been caused by aircraft encountering abrupt wind shear while attempting to take off or land. Quick positive action is required to counteract it. If you can anticipate the action that will be required, your response will be faster.

# CHAPTER 16

## LOW-LEVEL WIND SHEAR

### Section 1 - The Meteorology of Low-Level Wind Shear

**Introduction**

1. Wind shear was introduced in Chapter 11 in conjunction with boundary layer winds. The intent of the present chapter is to describe wind shear and its effects on take-offs and landings in greater detail and to provide instrument flight procedures to cope with it. This will be particularly pertinent for jet aircraft since they are the most seriously affected, but it will also be relevant for piston and turbo-prop aircraft.

2. To review briefly, you will recall that there are occasions when the wind through which an aircraft is flying changes so rapidly that it causes a change in the airspeed. Because of its mass, an aircraft has inertia that tends to prevent changes in its absolute velocity. When the wind changes at a rate greater than the aircraft can accelerate or decelerate because of its inertia, the airspeed changes.

3. Wind shears are described as increased performance shears or decreased performance shears. An increased performance shear occurs when an aircraft abruptly enters an increased headwind component or a decreased tailwind component. In this case the airspeed will increase. A decreased performance shear occurs when an aircraft abruptly enters a decreased headwind component or an increased tailwind component. In this case, the airspeed will decrease. After this abrupt change, the airspeed will gradually return to its original value. The ground speed changes too, but much more slowly than the airspeed.

4. Because the lift produced by a wing is proportional to the square of the airspeed, any changes in airspeed result in marked changes in the lift developed and, therefore, the vertical velocity of the aircraft. With an increased performance shear, lift increases. With a decreased performance shear, lift decreases.

5. There can also be crosswind shears that can cause the aircraft to weathercock into the new wind, and up and downdrafts that abruptly change the angle of attack of the wing. With a downdraft, the angle of attack will decrease, reducing the amount of lift being produced. With an updraft, the angle of attack will increase, and therefore increasing the amount of lift being produced. In extremely strong updrafts, or due to manoeuvring the aircraft in an updraft, the angle of attack may be increased to such an extent that a stall will occur.

**16** CHAPTER

## What it Does

6.  The effect of shear on an aircraft is particularly important during take-off or landing. What occurs will depend on the amount of shear, the altitude at which it is encountered, the characteristics of the aircraft and the reactions of the pilot. It is important to note that if shear is encountered in these phases of flight the aircraft will be at a low airspeed and high drag configuration with flaps and gear down. Shear is present during almost every take-off and landing but it can be so slight that you may have compensated for it without even realizing its presence. On the other hand, although it rarely occurs, the shear can be so strong that an aircraft cannot penetrate it safely. There have been several catastrophic accidents caused by wind shear.

7.  A crosswind shear will either shift the aircraft to the right or left of the desired path or weathercock it into the new wind. Continuing the approach in this situation requires sufficient altitude to make a heading correction to regain the desired approach track or heading.

8.  An airplane on approach is in a high drag configuration (gear and flaps down) and is normally flown at an airspeed just a little higher than that at which minimum drag for that configuration is produced. In Figure 16-1, the approach speed for an aircraft is 120 knots. If the speed should increase to 130 knots, the drag will increase and the aircraft will return to 120 knots. If the speed should fall to 110 knots, the drag will decrease, and again the aircraft will return to 120 knots. If the speed should drop further, however, to less than 100 knots, the drag will begin to increase. This latter situation can be aggravated by a decreased performance shear and the airspeed will decrease and the drag decrease further in a snowball effect which can quickly get beyond the recovery point unless the pilot takes prompt action.

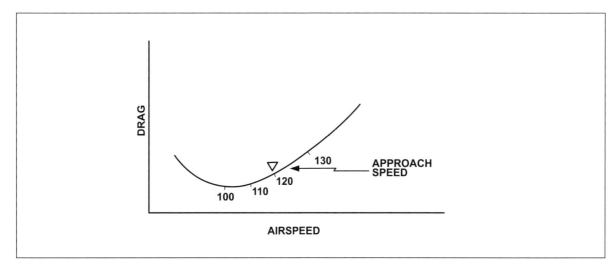

**Figure 16-1 Drag versus airspeed**

9.  Figures 16-2 and 16-3 illustrate decreased and increased performance shear situations. The shear layer in these examples is at a sufficiently high altitude on the approach that the various aircraft and pilot reactions are completed and the aircraft restabilized on the glide slope before touchdown. Note that, under actual conditions, the shear layer may be lower than that shown in the figures, and the wind may not be uniform under the shear layer. The situation as it exists at any particular instant will have to be dealt with without knowing precisely what lies ahead on the glide slope.

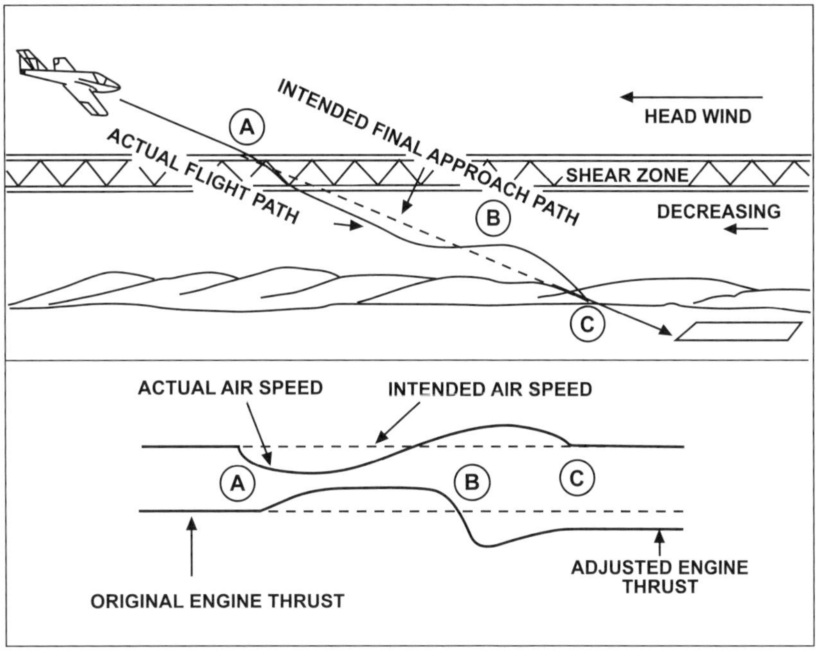

**Figure 16-2 Decreased performance shear (tailwind shear)**

10. Figure 16-2 depicts an aircraft on approach as it encounters a decreased performance shear. The top half of the figure shows the approach path and the wind shear and the bottom half, the airspeed and engine thrust as the aircraft passes along the approach path. The first instrument indication as the aircraft passes through the shear is an abrupt drop in airspeed. The loss of airspeed decreases the lift available so the rate of descent increases and the aircraft sinks below the glide slope (Point A). Should this occur at low levels, the pilot must react very quickly by applying power and raising the nose to regain the glide slope to avoid undershooting the runway. Note that if the aircraft is very low at this point, the nose cannot be lowered to regain the lost airspeed or else contact with the ground may occur. This and other flight procedures are explained in Section II of this chapter. As the aircraft is now in a decreased head wind, the extra power that was added will bring it above the glide slope (Point B) unless the pilot adjusts the throttles to something less than what was set originally (Point C).

11. Figure 16-3 shows an aircraft encountering an increased performance shear on approach. The first indication of the shear is an increased airspeed. As the airspeed increases, the rate of sink decreases and the aircraft rises above the glide slope (Point A). If this occurs very low on the approach, the aircraft will be high and fast and an overshoot might result. To regain the glide slope, the pilot must reduce power and lower the nose. Because of the spool-up time required for jet engines, caution must be used in jet aircraft in the amount of power that is reduced in case large amounts of power are later required on the approach. As the aircraft is now in a stronger head wind, the decreased power will drop it below the glide slope (Point B) unless the pilot adjusts the power to something more than he had on originally (Point C).

**16** CHAPTER

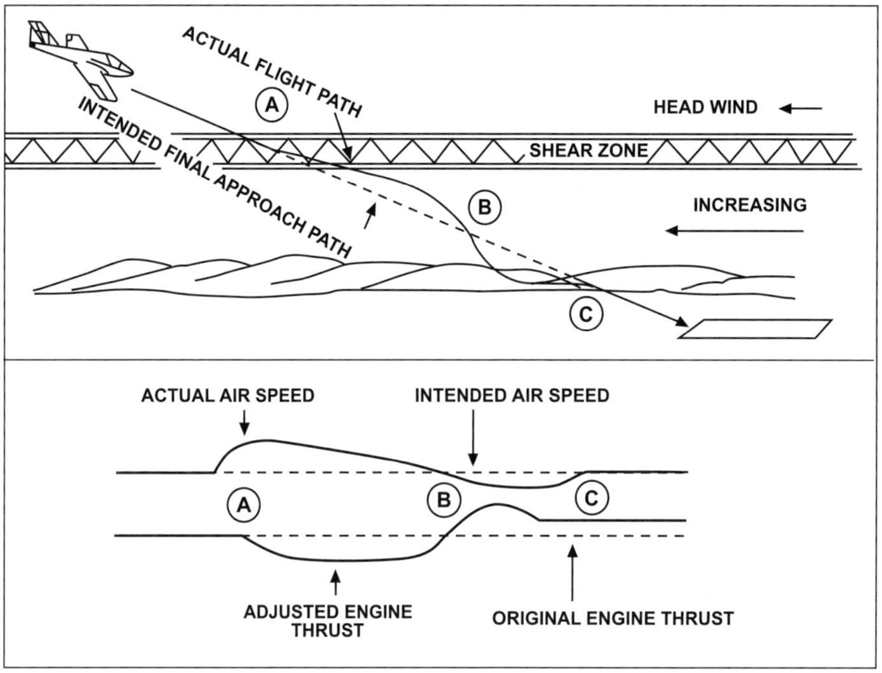

**Figure 16-3 Increased performance shear (headwind shear)**

12. Prompt and firm corrective action may be necessary to recover from a shear encounter so it is essential that you can quickly recognize when you are being affected by shear. Although advice as to whether a specific take-off or landing will encounter serious shear is not now generally available, meteorological conditions where shear is frequent are known. By understanding these, you will be alert for shear should it occur.

## Shear and Wind Gusts

13. Wind shears do occur with gusty winds that cause rapid fluctuations in airspeed and vertical velocity. These, however, are very transitory, so that any increases in headwind or tailwind components or vertical currents are almost immediately followed by decreases. This is totally different than the sustained shears that are being described in this chapter.

## Where Low-Level Wind Shear Occurs

14. Low-level shears can be particularly significant with:
   • Thunderstorms
   • The Rotary system of lee waves
   • Fronts
   • Low-level nocturnal jets
   • Valley, funnel and katabatic winds

   They can also occur with:
   • Eddies in the lee of hills or buildings
   • Convective thermals

## Thunderstorms

15. The flow of air in and around thunderstorms can produce strong shears and these have been the cause of most shear-related accidents. The diagram in Figure 16-4 demonstrates the basic nature of the problem. In any actual case, the airflow will normally be complicated by the presence of other thunderstorms, the meteorological situation and various topographical effects.

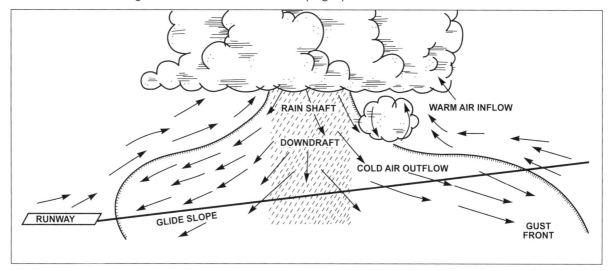

**Figure 16-4 Airflow under a thunderstorm**

16. The downdraft is a very local phenomenon, moving across the ground with the storm. Its diameter may be in the order of 1 to 3 miles, with the outflow spreading ahead of the storm from a depth of a few hundred to a few thousand feet and out to 5, 10, or possibly in rare cases, 20 miles ahead of the storm. The outflows from adjacent storms can merge. Because of this, conditions along the landing or take-off paths can vary from minute to minute so that even a PIREP from an aircraft immediately ahead may not indicate the situation that is to be encountered. PIREPs are still one of the best sources of information available on shear and should always be passed on when shear is encountered. Because of the very localized nature of the thunderstorm winds, the aerodrome anemometer which may be a mile away, may or may not indicate the actual approach or take-off winds. It can indicate a possible shear situation, however, if its wind is greatly different from that being encountered on approach.

17. The strengths of the downdraft and outflow are normally related to the vertical height of the thunderstorm. Downdrafts have been measured as high as 2,400 feet per minute and the thunderstorm gust front winds can reach destructive strengths of 100 knots. These winds diminish in intensity the further that they spread away from the downdraft centre. Not all thunderstorms develop hazardous shears or downdrafts, but all thunderstorms do have shears and downdrafts present to some extent.

18. The winds, shears and vertical drafts encountered on an approach or take-off will depend on the location of the downdraft. The situation where the downdraft is centred on the glide slope during an approach is presented in Figure 16-5. The top of the figure indicates the airflow in relation to the glide slope and the path that the aircraft will take as it encounters shears and vertical drafts. The bottom of the figure indicates the changes in airspeed, rate of descent and thrust at various points along the glide slope. The pilot is trying to maintain the glide slope by reacting to the information provided by the instruments and is unaware of the situation along the glide slope ahead.

**16** CHAPTER

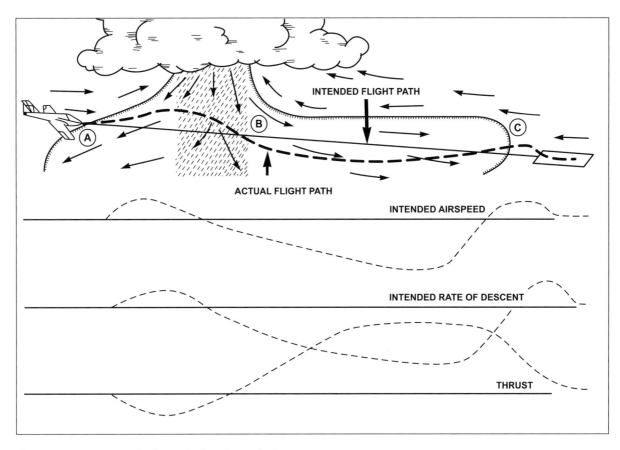

**Figure 16-5 Approach through the downdraft**

19. On this approach, the aircraft will pass from a light tail wind associated with the inflow through an increased performance shear as it encounters the outflow of the downdraft (Point A). This will cause an increase in airspeed and an increase in lift so the aircraft will tend to rise above the glide slope. To counteract this, thrust will have to be reduced and the nose of the aircraft lowered. The aircraft now enters the downdraft and then a decreased performance shear (Point B). This will cause a serious drop in airspeed and a marked increase in the rate of descent so that the aircraft will tend to sink below the glide slope. This is the danger area. The aircraft is now low and slow. Thrust must be increased quickly and the nose lifted to regain the glide slope. Increased performance shear will next be encountered (Point C) as the aircraft passes through the gust front and encounters a head wind). The aircraft will then become high and fast and corrections may have to be made to prevent overshooting.

20. Accidents have also occurred on take-off when the downdraft has been in the take-off path. Aircraft have taken off into the cold air outflow, entered the downdraft at very low speed and little height, and settled onto the ground as airspeed was lost further due to the tailwind-shear on the other side of the downdraft. Figure 16-6 illustrates this situation.

21. The aircraft commences its take-off with a head wind (Point A). It encounters a downdraft (Point B) and then passes into a decreased performance shear (Point C). The resulting loss of airspeed and lift may cause it to settle onto the ground. Should it maintain flight, it will enter an increased performance shear zone (Point D) and be out of trouble.

22. Downdrafts have also developed from very high-based thunderstorms from which virga (rain that evaporates before it reaches the ground) is falling. If the air beneath the storm is excessively dry such as can occur over the Prairies, the very rapid evaporation of the rain can cool the air to an extent that it plummets earthward as a strong cold draft.

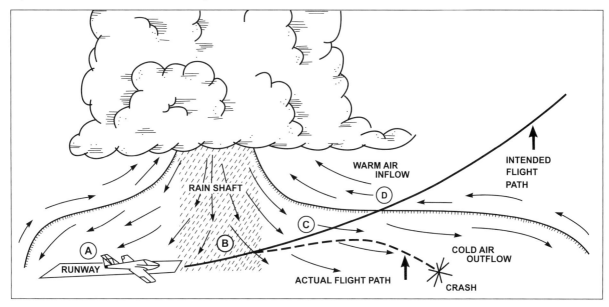

**Figure 16-6 Take-off through the downdraft**

23. Terms now frequently used in describing particularly hazardous downdrafts are as follows:

    a. A "Downburst" is a strong downdraft inducing an outward burst of damaging winds on or near the surface.

    b. A "Microburst" is a small, more intense downdraft embedded within a downburst. The effects on flight are the same as those described for a downburst.

    c. An "Outburst" is the quasi-horizontal flow of air away from the downburst.

    d. An "Outburst Centre" is the centre of the downburst where the vertical air current hits the surface and spreads out violently.

24. While an outburst can extend for a considerable distance away from the outburst centre and create unexpected shear on this account, it is the microburst with its small-scale, severe shear and downdraft that is the most hazardous. They can be less than a mile across; so small as to be difficult to detect.

## Effect of Rain

25. When an aircraft flies through the very intense rain of a thunderstorm rain shaft, its wings become coated with water. There is some evidence that this coating may roughen the wing surface sufficiently to increase the stalling speed. This may have been an added factor in some of the accidents caused by wind shear.

## Lee Wave Rotors

26. Figure 16-7 illustrates the wind flow in relation to a lee wave rotor. The situation depicted is a decreased performance shear. There may be visual evidence that shear is probable from the presence of a rotor cloud overhead and dust being picked up and blown in the direction of flight. The aircraft may enter turbulence as the shear is encountered. The rotor nearest the ridge (primary rotor) normally will have the strongest shears so that aerodromes near the base of ridges are particularly prone to this problem.

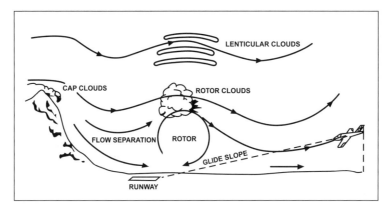

**Figure 16-7 Lee wave rotor shear**

## Frontal Surfaces

27. Frontal wind shear was described in Chapter 11. It is essential to realize that frontal slopes are extremely shallow so that take-off and landing paths will cut through them. Considering typical frontal slopes and speeds, shear at a terminal will normally be significant for take-offs and landings for something less than an hour for a cold front and something around two or three hours for a warm front.

## Low-Level Nocturnal Jets

28. You will recall from Chapter 11 that low-level jets are at maximum strength a few hours after midnight and form mainly over the Prairies when skies are clear. A report of light or calm surface winds while you are correcting for strong winds on the approach at around 1,500 or 1,000 feet is an obvious clue that you are due for a shear encounter.

## Other Shear Situations

29. There are a few other meteorological conditions of a localized nature that can produce wind shear that might create difficulties during take-off or landing. The pull of gravity can cause a surface layer of cold, and, therefore, dense air to flow down slopes or valleys or out through mountain gaps. These are "Valley", "Katabatic" and " Funnel Winds." They can be very strong and produce intense shears. They are particularly relevant at Arctic terminals or along mountainous coasts.

30. Of a more minor nature, shears can occur because of eddies in the lee of hangars of buildings on an airport and because of convective thermals. "Thermals" are bubbles or shafts of rising warm air that create an inflow. This could produce minor shears as well as small vertical updraft on approach or take-off.

## Section 2 - Flight Procedures for Low-Level Wind Shear

This material has been developed in conjunction with the Division Instrument Check Pilot and the Canadian Forces Instrument Check Pilot School.

### Background

31. Wind shear is not a new phenomenon. Since the beginning of aviation, aircraft have been encountering wind shear and pilots have been coping with its effects. Modern jet aircraft with sweptback wings, high take-off, approach and vertical speeds and large payloads are more susceptible to shear than are older piston, propeller driven, aircraft. This susceptibility has resulted in several serious accidents in the last few years. The very detailed studies that were made during the accident investigations have resulted in a considerable body of information regarding conditions causing shear and the flight procedures that are required to cope with it. Unfortunately, the information is not yet complete and more must be learned both about the shears and the flying procedures required to cope with them before the problem can be considered solved.

32. The description of the flight procedures that follows will provide general techniques that should assist you if you encounter shear. Your aircraft's AOIs should also be consulted. You should be aware that, although critical shear occurrences are rare, shear beyond the capability of an aircraft to recover can occur.

### Know Your Aircraft

33. There is ample evidence to indicate that accidents have occurred because the pilot did not use the full capability of his aircraft. Crashes due to shear normally occur because the aircraft sinks onto the ground. The techniques that have evolved are designed to prevent this by adjusting flight procedures to prevent or arrest the sink. To do this you must know your aircraft thoroughly. You should be familiar with the:

- Variation of landing and take-off distances with different surface winds, flap settings and airspeed.

- Normal acceleration as indicated by the airspeed indicator during the take-off roll.

- Airspeed to get the maximum rate of climb for different thrust and flap settings.

- Normal pitch attitude during climb and letdown.

- The difference between pitch and angle of attack.

- Spool-up time required by your engines.

- Speed range between stick shaker and stall.

- Airspeed at which INS or Doppler no longer gives wind or ground speed information.

- Manoeuvrability of your aircraft at slow speeds.

- Acceleration capability of your aircraft with pitch and with throttle.

- Amount of height that can be gained by bleeding off airspeed and the amount of height lost by increasing airspeed with pitch.

- Sinking feeling called "Kinesthesis" that occurs when you encounter subsiding air.

## Energy Management

34. An aircraft in flight has kinetic energy due to its motion, and potential energy due to its height above ground.

    Kinetic Energy: $KE = 1/2MV^2$
    Potential Energy: $PE = Wh$

    where M is the mass of the aircraft, V the velocity, W the weight of the aircraft and h the height if the aircraft is above ground. The velocity of the aircraft is particularly important because the kinetic energy varies as the square of the velocity. The aircraft has kinetic energy due to its horizontal motion through the air as indicated by the airspeed indicator and also due to its vertical motion as indicated by the vertical speed indicator.

35. The horizontal kinetic energy can be exchanged for potential energy by pitching the nose of the aircraft up, thus exchanging the airspeed for height. Potential energy can be exchanged for kinetic energy by pitching the nose of the aircraft down and exchanging height for airspeed. The height gained or lost is independent of the mass of the aircraft. The exchange of airspeed for height can be used in an emergency to temporarily increase the rate of climb or decrease the rate of sink. The table in Figure 16-8 indicates the amount of height that can be gained by exchanging airspeed for height. Note that the higher the airspeed, the greater the height gain. A total change of energy (KE + PE) can only be obtained by use of throttle or from vertical air currents.

| 10-Knot Change | | Equivalent | 20-Knot Change | | Equivalent |
| --- | --- | --- | --- | --- | --- |
| **From** | **To** | **Altitude, Ft.** | **From** | **To** | **Altitude, Ft** |
| 150 | 140 | 128 | 150 | 130 | 247 |
| 140 | 130 | 119 | 140 | 120 | 230 |
| 130 | 120 | 111 | 130 | 110 | 212 |
| 120 | 110 | 102 | 120 | 100 | 195 |
| 110 | 100 | 93 | 110 | 90 | 177 |

**Figure 16-8 Table of height gained with changes in airspeed**

36. If you have been caught in a decreased performance shear and have lost airspeed even though you are at maximum power, you could lower the nose and regain the airspeed by trading potential energy for kinetic energy. This, however, either increases the sink rate or decreases the rate of climb and compounds the problem of maintaining height caused by the decreasing performance shear. At very low levels, it would be preferable to maintain the lower speed than to push the nose over and risk ground contact. A full instrument scan is essential in shear so vertical speed and altitude must be watched as closely as airspeed.

CHAPTER **16**

## Angle of Attack and Pitch Attitude

37. Figure 16-9 reviews the relationship between the angle of attack ( $\propto$ ) and the pitch attitude. The "Angle of Attack" is the angle between the wing chord and the relative airflow, and the "Pitch Attitude" is the angle between the longitudinal axes of the aircraft and the horizontal.

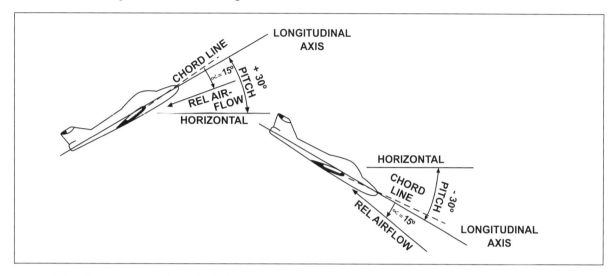

**Figure 16-9 Angle of attack and pitch attitude**

38. Figure 16-10 illustrates the behaviour of an aircraft on entering a downdraft with the pitch attitude held constant. At (a) the aircraft is flying straight and level with an angle of attack. At (b) the relative airflow abruptly acquires a downward component because of the downburst. Because of its inertia, the aircraft takes a considerable time to acquire the downward motion of the downdraft so the angle of attack decreases and the amount of lift produced by the wing decreases. This change is not apparent in the pitch attitude. At (c) the aircraft has acquired the motion of the downdraft and the angle of attack has returned to what it was originally. This latter situation may not occur if the aircraft is in the downdraft for only a very short period.

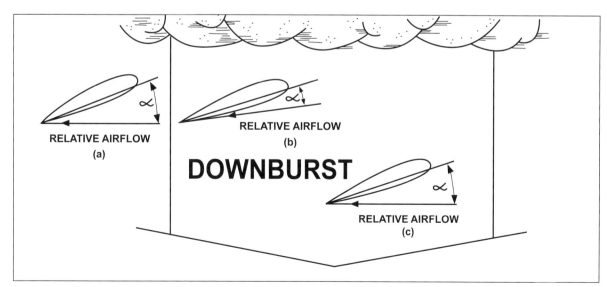

**Figure 16-10 Changes in the angle of attack in a downburst**

39. To compensate for the decrease in the angle of attack on entering a downburst you must place the aircraft in an abnormally high pitch attitude. When you fly out of the downburst and particularly if you should then enter a decreased performance shear area, you must be prepared to quickly reduce the pitch angle to that required for the new situation.

## Recognition of Wind Shear

40. Wind shear encounters last for only a matter of seconds. For a successful penetration it may be necessary to react almost instantaneously. If you are aware ahead of time that you are going to encounter shear, and whether it will be an increasing performance shear or a decreasing performance shear, your reaction time will be less.

41. Chapter 11, "Boundary Layer Winds and Turbulence," and Section 1 of Chapter 16 have provided considerable detail of the meteorological and topographical conditions that can cause shear. You should be able to relate all of this information to the approach or take-off that you are about to make.

42. Prior to flying, you should assess the possibility of shear at your weather briefing, as well as check for the numerous shear-causing phenomena that have been covered in this manual there is some direct information that may be available. Shear is occasionally, although not often, mentioned in terminal forecasts and in weather reports. Upper wind measurements made by "Pilot Ballons" (pibals) or radiosondes may indicate strong shear. Forecast upper winds however, are not useful because the vertical distance between forecast levels is too great. PIREPs may be available and are the best source of information.

43. In considering shear situations, you should be aware that, in the case of thunderstorms, the shear is extremely transitory, changing from minute to minute, whereas shears due to inversions, fronts or topography are much more persistent and may last for hours.

44. There are many visual clues to shear, which may be evident either prior to take-off or during an approach. Shear is common at the top of inversions. Stratus cloud is topped by an inversion, fog forms under an inversion and smoke flattens out at the top of an inversion. An approaching gust front may be evident from blowing dust and debris; and/or lee wave rotors by blowing dust or snow and a rotor cloud. The rain shaft of thunderstorms or virga from high-based thunderstorms may be seen visually or by radar.

45. A comparison of the wind at some point on the approach with that on the ground will give an indication of the general shear throughout the approach (Figure 16-11). The crosswind component seldom has serious consequences so a comparison of the head or tailwind component is sufficient. The surface wind that is used for this must be current, an ATIS wind, for example, is unsatisfactory. Caution must be used if thunderstorms are in the vicinity because the small size of downbursts and microbursts can cause the threshold wind to be much different from the wind obtained at the anemometer site.

46. Some aircraft are instrumented to provide in-flight wind information. If the equipment provides this information down to low altitudes and low airspeeds, it can be used to compare the in-flight wind to the surface wind throughout the approach and the shear can be continuously calculated. Other aircraft have equipment that provides ground speed so that this can be used to calculate the head or tailwind component being encountered and this can be compared to the surface head or tail wind

component. Even without special equipment, the ground speed can be easily calculated on a precision approach.

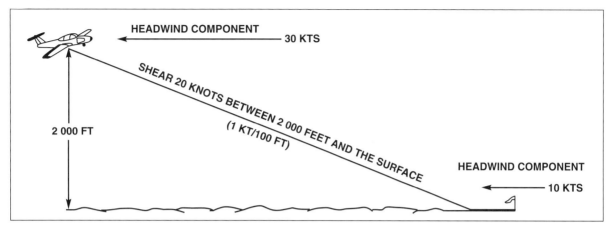

**Figure 16-11 Average shear along the approach**

## Computing Ground Speed on a Precision Approach

47. Once the aircraft is stabilized on the approach and before the pilot workload gets too high, an estimate of the ground speed can be obtained as follows:

For a 3° glide slope $\qquad GS = \dfrac{VSI}{5} - 10$

For a 2 1/2° glide slope $\qquad GS = \dfrac{VSI}{5} + 20$

(GS is ground speed in knots, VSI is vertical speed in feet per minute. The headwind or tailwind component can then be calculated using the indicated airspeed.)

48. The ground speed can also be obtained for any precision glide slope using a flight computer such as the Dalton Type MB-4A or Type CPU-26A/P:

   a. Place the glide slope in the drift correction window.

   b. Read the ground speed in knots off the minutes scale opposite the rate of descent on the miles scale.

49. For Dalton Computers that do not have a drift correction window, and for Jeppesen computers, the following can be done: Mark an index on the miles scale at 32 miles for a 3° glide slope and 26.4 miles for a 2 1/2° glide slope. Set the time index at the appropriate glide slope index and read the ground speed off the minutes scale opposite the rate of descent on the miles scale.

50. Figure 16-12 illustrates how the ground speeds obtained from the computers, or from tables, could be placed directly on the vertical speed indicators where they could be continuously referred to during an approach.

**16** CHAPTER

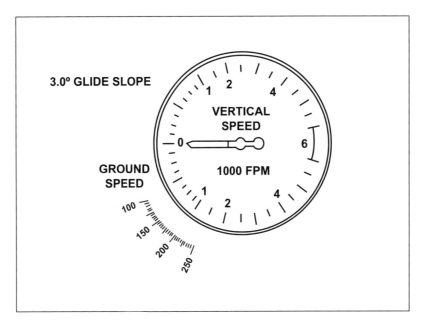

**Figure 16-12 Vertical speed indicator adapted to read ground speed**

51. An indication of the amount of shear that might be encountered can be obtained from the following table, which describes the changes of wind in the vertical.

| Light | 0-4 Knots/100 feet |
| Moderate | 5-8 Knots/100 feet |
| Strong | 9-12 Knots/100 feet |
| Severe | over 12 Knots/100 feet |

**Figure 16-13 Table of wind shear intensity**

52. Aircraft vary in their ability to cope with shear. Several countries use 8 knots/100 feet as the maximum wind shear that an automatic landing system must handle for certification. This intensity of shear is fairly often exceeded under certain atmospheric conditions. Shears of 3 to 5 knots per 100 feet are common and are not hazardous. With shears of 9 knots per 100 feet considerable difficulty might be encountered so that, from this point on, a missed approach may be required.

53. Because the rate of change of the wind being encountered is the most important factor, aircraft with high approach speeds are more seriously affected because they pass through the shear layer more quickly. The table below indicates the time it would take for an aircraft to pass through a 20-knot shear in 100 feet along a 3° glide slope at different approach speeds. The rate of shear is indicated in the last column.

| Indicated Airspeed | Time | Rate |
|:---:|:---:|:---:|
| 60 kts | 20 seconds | 1 kt/sec |
| 120 kts | 10 seconds | 2 kt/sec |
| 150 kts | 8 seconds | 2.5 kts/sec |

**Figure 16-14 Rate of shear table at different airspeeds**

54. To counteract a shear of 2.5 knots per second by reducing the pitch angle in a typical jet liner using take-off thrust would reduce the rate of climb to zero. If a downdraft should be encountered at this time, the results could be disastrous.

55. If shear is suspected, the ground speed could be computed or measured as soon as you are stabilized on the glide slope. This would possibly be at around 1,500 or 2,000 feet. By comparing the wind obtained at this level with the surface wind, the total shear on your approach will be known, as well as whether it is a decreasing or increasing performance shear. At this point you will not know if the shear is concentrated in a shallow and therefore dangerous layer or if it occurs in a deep but not dangerous layer (Figure 16-15).

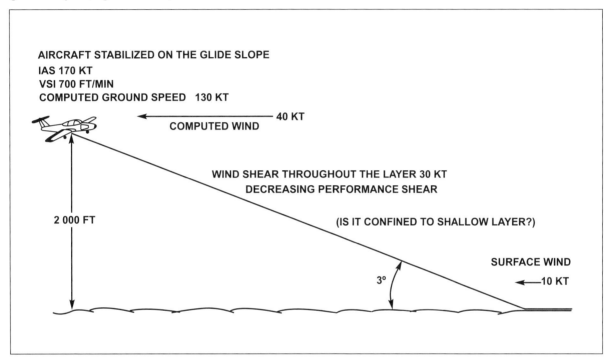

**Figure 16-15 Calculation of the average shear throughout a precision approach**

56. Figure 16-16 illustrates how you can assess your situation as you progress along the glide slope. A rule of thumb is that a change of vertical speed by 50 feet per minute is required for a 10-knot wind change for either 2 1/2° or 3° glide path. The vertical speed becomes less with increasing winds and more in decreasing winds. You have stabilized your approach on the glide slope at 2,000 feet and have calculated your ground speed and from this obtained a headwind component of 40 knots. The surface headwind component is 10 knots. The shear is 30 knots in 2,000 feet or 1.5 knots per 100 feet decreasing performance shear, which is insignificant. No thrust adjustment has been required by the time you reach 1,000 feet. This means that the headwind component has remained unchanged at 40 knots. The shear for the remainder of the approach is now 30 knots in 1,000 feet or 3 knots in 100 feet decreasing performance shear which is still insignificant. As you continue the approach you find it necessary to reduce thrust so that the vertical speed increases by 50 feet per minute by 500 feet. This implies that the wind has become lighter by 10 knots. If any correction is required, it will be for a decreased performance shear and will require the addition of thrust and a pitch-up to counteract the loss of airspeed and sink that will result.

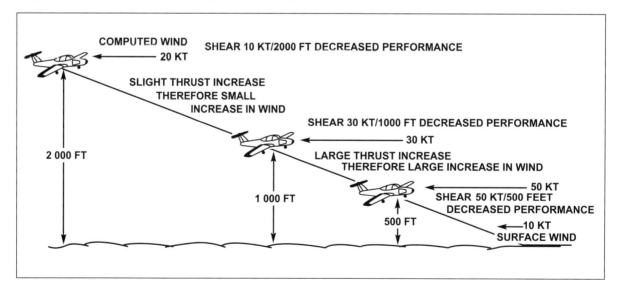

COMPUTED WIND    SHEAR 30 KT/2000 FT DECREASED PERFORMANCE
← 40 KT

NO THRUST ADJUSTMENT REQUIRED
THEREFORE NO CHANGE IN WIND

SHEAR 30 KT/1000 FT DECREASED PERFORMANCE
← 40 KT
REDUCED THRUST
THEREFORE
LIGHTER WIND

2 000 FT

SHEAR 20 KT/500 FEET
DECREASED PERFORMANCE
← 30 KT

1 000 FT

500 FT

SURFACE WIND
← 10 KT

**Figure 16-16 An improving shear situation**

57. Figure 16-17 is an example of a worsening shear situation such as would occur with a low level jet stream. In this case, you are again stabilized at 2,000 feet and have calculated a headwind component of 20 knots which gives a decreased performance shear of only 10 knots throughout the approach. As you descend you find that you must add a small amount of thrust to maintain the glide slope. The vertical speed decreases by 50 feet per minute. This implies a slightly increasing head wind so that by 1,000 feet the decreased performance shear is 30 knots per 1,000 feet. As you continue down, you find that you require a large thrust increase with the vertical speed decreasing by a further 100 feet per minute. The wind therefore is increasing markedly and the decreased performance shear is around 50 knots in 500 feet. You should be prepared for a large drop in airspeed and a large increase in sink and possibly a missed approach.

COMPUTED WIND    SHEAR 10 KT/2000 FT DECREASED PERFORMANCE
← 20 KT

SLIGHT THRUST INCREASE
THEREFORE SMALL
INCREASE IN WIND

SHEAR 30 KT/1000 FT DECREASED PERFORMANCE
← 30 KT
LARGE THRUST INCREASE
THEREFORE LARGE INCREASE IN WIND

2 000 FT

← 50 KT
SHEAR 50 KT/500 FEET
DECREASED PERFORMANCE

1 000 FT

500 FT

← 10 KT
SURFACE WIND

**Figure 16-17 A worsening shear situation**

## Approach Techniques

58. Very extensive experimentation has been done on various flight techniques for all sizes of transport aircraft. Pilot opinion of these tests indicates that a ground speed/airspeed comparison is the best aid in assessing shear during an approach. A ground speed read-out of some type is required. In this procedure you should select a minimum ground speed by subtracting the headwind component of the surface wind from the minimum airspeed that you will accept on approach. You should then fly the aircraft so that neither ground speed nor airspeed is allowed to drop below these minimums.

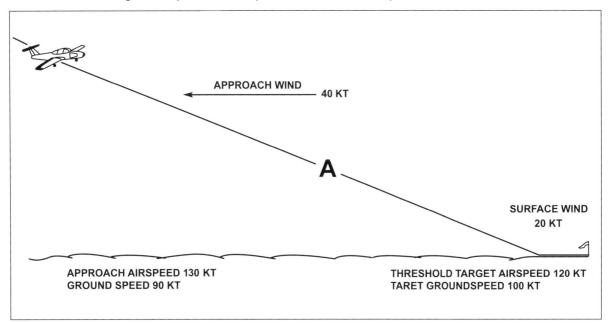

**APPROACH WIND** 40 KT

**A**

**SURFACE WIND 20 KT**

APPROACH AIRSPEED 130 KT
GROUND SPEED 90 KT

THRESHOLD TARGET AIRSPEED 120 KT
TARET GROUNDSPEED 100 KT

**Figure 16-18 Minimum ground speed-airspeed technique for flying shear**

59. Figure 16-18 illustrates the benefit of using this system. Remember that, in shear, the airspeed changes rapidly but the ground speed changes slowly. As indicated in the figure, if you cross the threshold with a ground speed of 100 knots, you will have your targeted airspeed. If at point A you should have let the ground speed drop to 90 knots because of the strong head winds, the airspeed will drop below your target airspeed when you enter the shear zone ahead of you. As you can see, this system is particularly useful because it predicts the shear ahead of you so that a missed approach can be commenced before the shear is encountered. It does depend on having a correct threshold wind, however, which may not be available in a downburst situation.

60. With a normal headwind approach, if the difference between the airspeed and the ground speed is increasing, you are in an increasing performance shear. If the difference is decreasing, you are in a decreasing performance shear.

61. There are several other techniques that can be considered. These are general in nature and your AOIs must also be consulted.

**16** CHAPTER

a. Use a lesser flap setting. This should allow a better acceleration capability by reducing drag. Increased touchdown speed, longer ground rolls and a potential for hot brakes must, however, be considered as trade-offs when using this technique.

b. Consider using an airspeed pad if a decreasing performance shear is anticipated. Remember that airspeed is energy in the bank. Your three sources of energy are:

   (1) Engines - they may be slow to accelerate (spool-up time 8-12 seconds for straight jets, longer for fanjets).

   (2) Altitudes - it may be traded for airspeed, but this increases the sink rate and at low level altitude is not available.

   (3) Airspeed - the only one that is instantly available. The Boeing Aircraft Company recommends using a stabilized approach speed of no more than 20 knots above VREF (the normal approach speed computed for the landing weight) below 500 feet AGL for transport-type aircraft.

c. Determine what approach limits you will accept during your approach planning. If these limits are exceeded go around. Some considerations are:

   (1) The approximate power settings (EPR, RPM, Fuel Flow) are to be stabilized as far out on final approach as possible. Avoid large power reduction.

   (2) The number of knots of airspeed that are to be added on the final approach (decreasing performance shear).

   (3) How much of an airspeed loss will be accepted.

   (4) Vertical velocity - Remember that kinetic energy also acts in the vertical plane. Any large amount of vertical kinetic energy must be dissipated before touchdown.

d. Cross-check all available instruments down to touchdown. Deviations can be recognized sooner by monitoring the flight instruments than by peering out a rain-splattered windscreen.

e. Don't accept large glide slope deviations before taking it around.

f. Avoid trimming at low altitude. Any need for large amounts of trim indicates a changing airspeed.

g. Do not make large power reductions until beginning to flare. Spool-up times can be critical if a sudden shear is encountered at low altitude.

h. Use VASI lights to keep you in approach, particularly in close when visual. Remember that VASI lights do not give trend information, they only tell you if you are high or low.

i. Pick your aim point on the runway and use it. Do not "duck under" or change your aim point because you might encounter a decreasing performance shear and land short.

62. Both Boeing and Lockheed aircraft companies have conducted performance studies and both have reached the following conclusions:

   a. If a severe decreasing performance wind shear and/or downdraft is encountered, rated take-off thrust should be applied and the airplane pitched up to at least the go-around attitude.

   b. If this does not arrest the descent, the pitch attitude should be further increased slowly to exchange airspeed for climb capability until further altitude loss can be prevented. (Be prepared to fly to the stick shaker and add full power.) No attempt should be made to accelerate back to the approach speed until the shear has been penetrated. Care must be taken to have some reserve airspeed for a flare, should ground contact be inevitable.

## Take-off Techniques

63. Shear is almost always present in the low levels of the atmosphere and it is difficult to know whether it will be hazardous or not. Some situations are fairly obvious. For example you should never take off directly towards a thunderstorm. Pilot reports may indicate that hazardous shear is occurring so it would be prudent to wait until the situation improves. On the other hand, there may be evidence that although there is significant shear, it can be traversed safely. You should be aware that the hazards and problems associated with take-offs made in severe shear and/or downdrafts are similar to those encountered during approaches. The hazard will arise either because the maximum performance of the aircraft is insufficient or the pilot is not getting the maximum performance out of the aircraft. The latter has occurred because of the pilot's concern about abnormally low airspeed in a decreasing performance shear and the subsequent degradation of climb capability resulting from the pilot's attempts to rapidly regain speed by reducing pitch attitude.

64. The following take-off techniques are suggested:

   a. A full panel scan during initial climb-out. This is of the utmost importance.

   b. Fly at a high pitch attitude if necessary. Speeds down to stick shaker can still give a large rate of climb. No attempt should be made to accelerate back to the initial climb speed until the shear has positively been traversed. Attempts to accelerate will significantly reduce climb capability.

   c. Don't hesitate to use all the available thrust if it hasn't already been set.

   d. If practical, use the longest runway available.

   e. Consider using full take-off power as opposed to a reduced thrust take-off.

65. Decreasing performance shear or a downdraft may be encountered during the take-off run. Should this occur, the following are suggested:

**16** CHAPTER

a.  Use a higher than normal lift-off attitude to unstick the aircraft. Shear may be evident by abnormally slow acceleration as indicated by the airspeed indicator during the take-off roll.

b.  On initial climb-out do not accelerate back to normal climb speed. A high angle of attack may be required.

c.  Do not make any stabilizer trim inputs until stabilized conditions have again been established.

d.  Be prepared for penetration of additional areas of horizontal shear or downdrafts.

## Pilot Reports

66.  Since ground-based instruments for measuring low-level wind shear are not normally available, pilot reports (PIREPS) are essential. If wind measuring equipment is available in the aircraft, the wind and altitude both above and below the shear layer should be provided. If you do not have this equipment, you should indicate the loss or gain of airspeed and the altitude at which it occurred. This information can be amplified with other pertinent remarks such as "maximum thrust required" or "severe turbulence."

67.  You should try and obtain pilot reports as an additional source of information to base your decisions on. They must be used with caution, however, and you should not allow others to make decisions for you. Particularly with thunderstorms, conditions change very rapidly. You should not expect to encounter conditions identical to those reported by the pilot just ahead of you.

68.  Everyone has flown an approach where, for some unknown reason, the aircraft deviated from the glide path. That bump or dip was most likely caused by a wind shear. Wind shear is a relatively new frontier and its effect on high performance aircraft has just recently been understood. It is hoped that someday instrumentation both on the ground and in the cockpit will be available to warn us of impending wind shear. Until that time, we must use everything available. Pilot reports, weather conditions and comparisons of approach ground speed with expected threshold ground speed are the tools available. You must know the weather conditions that may produce substantial wind shear. Finally, you must know what to do in order to recognize and cope with the shear once it is encountered.

# Summary - Chapter 16

- **Meteorological Factors**

- Shear causes a change in the airspeed and amount of lift.

- With an increased performance shear, airspeed and lift increase.

- An increased performance shear occurs with a rapid increase in a head wind, or a rapid decrease in a tail wind.

- With a decreased performance shear, airspeed and lift decrease.

- A decreased performance shear occurs with a rapid decrease in a head wind, or a rapid increase in a tail wind.

- An increased performance shear will cause an aircraft to rise above the glide slope. Corrective action may cause the aircraft to go below the glide slope.

- A decreased performance shear will cause an aircraft to sink below the glide slope. Corrective action may cause the aircraft to rise above the glide slope. A stall is possible.

- Thunderstorm Shear - The shear is associated with the outflow and gust front and can extend several miles from the downdraft centre.

- The effect of the downdraft can be added to a decreased performance shear to increase the hazard.

- There is some evidence that roughening of the wing surface by heavy rain can increase the stall speed.

- Rotor Shear - The primary rotor is most severe.

- The rotor cloud and lifting of surface dust may provide visual evidence of the shear area.

- The aircraft may encounter turbulence that will indicate shear may be encountered.

- Frontal Shear - The frontal surface must be at a low level over the airfield to be hazardous.

- The shear occurs with warm fronts and fast moving cold fronts.

- The low-level jet, katabatic, valley and funnel winds can all produce hazardous shear.

- Minor shear can occur in the lee of obstructions and because of convection.

**Summary - Chapter 16 continued...**

- **Flight Procedures**

- You must be thoroughly familiar with the performance of your aircraft.

- Accidents occur because the aircraft sinks onto the ground.

- Airspeed can be exchanged for height in a decreasing performance shear if necessary.

- Pitching down to regain airspeed in a decreasing shear may cause the aircraft to strike the ground. It is better to fly at the reduced airspeed until the shear is cleared.

- Entry into a downburst reduces the angle of attack so that an abnormally high pitch attitude is required.

- You must be able to relate the various meteorological factors that cause shear to the approach or take-off that you are about to make and check for these at your weather briefing.

- There frequently are visual clues that shear is present.

- During an approach the surface wind speed should be continuously compared with the wind being encountered. The approach wind can be assessed by:

  - aircraft instruments if they are available,
  - computing ground speed on a precision approach,
  - comparing the rate of descent with a normal rate of descent.

- If instrumentation is available, a recommended approach technique is to select a minimum ground speed and a minimum airspeed and to initiate a missed approach if either of these is reached.

- Other landing techniques include:

  - the use of less flap,
  - an airspeed pad,
  - careful cross-check of all instruments down to touchdown,
  - executing a missed approach if large glideslope deviations occur,
  - not trimming at low levels,
  - not making large power reductions at low levels,
  - using VASI lights,
  - picking an aim point on the runway and not undershooting it,
  - use of take-off thrust and go-around pitch attitudes,
  - exchanging airspeed for climb capability down to stick shaker if necessary.

- Take-off techniques include:

  - a full panel scan during initial climb-out,
  - staying at a high pitch attitude if necessary,
  - use of all available thrust,
  - use of the longest runway.

- An encounter with decreasing performance shear on the runway requires:

  - a higher than normal lift-off attitude,
  - not accelerating to normal climb speed,
  - not trimming until the shear has been transmitted.

- Pilot Reports are essential. Use them and make them available to other pilots.

- QUICK RECOGNITION OF A SHEAR ENCOUNTER AND RAPID CORRECTIVE ACTION IS NECESSARY.

**16** CHAPTER

# Chapter
# 17

During an approach, the weather obviously affects the decision whether to complete the landing or carry out a missed approach. For a correct decision, it is important to relate the available and visible weather information to its effect on an approach.

# CHAPTER 17

## WEATHER FACTORS DURING TRANSITION

### Introduction

1. Whether the transition from instrument to visual flight during the final phases of an approach is successfully carried out or not is determined by several factors. These include weather conditions, type of runway landing aids, aircraft type and pilot techniques. The transition is greatly assisted if what is ahead can be anticipated. A knowledge of the weather information that is provided to you, and an understanding of the effects that the low-level weather will have, help during the transition.

### Cloud Bases

2. The cloud base, particularly stratus, is not clean cut and flat, but ragged, tenuous and undulating. The actual base is therefore indeterminate and difficult to state as a specific height. What is provided is a height obtained by measuring the cloud base with a ceilometer. The type of ceilometer currently in use is based on an optical radar principle using eye-safe laser.

3. One problem with the laser ceilometer is to provide an accurate height during periods of surface obscuration or precipitation. In this situation, the height or the vertical visibility has to be confirmed by an observer or by software in the case of an automated station.

4. As you reach the reported cloud height on your approach, you will be looking ahead through the tenuous base and may not be able to make immediate visual contact with the runway environment. Terms such as "Ceiling Ragged" or "Ceiling Diffuse" in the weather report should alert you that such conditions exist.

5. Even without a ceiling, the cockpit visibility can be seriously restricted because of ragged cloud filaments in the approach area. Pilot reports indicating the height at which the runway became visible are of the greatest benefit, and pilots are encouraged to pass along this information.

6. The cloud base can also be undulating so that the height changes constantly and is different at the break-out point than at the observing site. When a ceiling is below 3,000 feet and its height is fluctuating by one quarter or more of its mean value, the range of variation will be indicated in the remark group of the METAR. For example, the METAR might report a ceiling at 400 feet but will mention that it is between 300 and 500 feet in the remark group. In this example, patches of cloud restricting vision down to 300 feet or slightly lower can be anticipated.

7. Cloud heights of up to 10,000 feet are rounded off to the nearest 100-foot level. This means, for example, that ceilings measured at 160 feet and 250 feet will both be reported as 200 feet. With low cloud the actual measured height may be reported in the remark group of the METAR.

8. Since cloud bases are not necessarily at a uniform height, cloud over different runways may be at slightly different heights than at the observation site. Some aerodromes are known for having lower

**17** CHAPTER

cloud bases on some approaches. This type of local knowledge can frequently be obtained from the weather office or from pilots familiar with the terminal.

## Obscured and Partially Obscured Conditions

9.   While there is difficulty in providing an exact height for many cloud bases, the problem is much greater for obscured conditions. An obscured condition is said to exist when the sky is not visible because of such things as snow or fog. In this case, the vertical visibility into the obscuring medium is considered to be the ceiling. Since there are no objects to be seen to assess this vertical visibility, it is extremely difficult to state what it is.

10.   The line of sight from the cockpit will be through the obscuring medium all the way to the runway, so a considerable reduction of visibility will result.

11.   In a partially obscured condition, some sky is visible through the obscuration so no ceiling height can be given. However, cockpit visibility will still be through the obscuring medium and will be reduced.

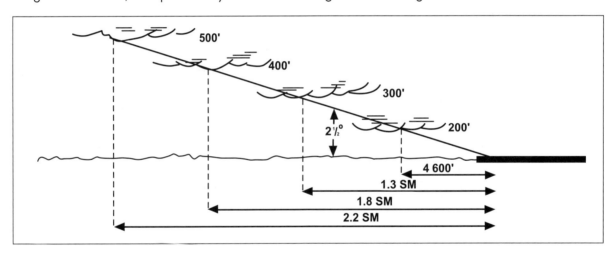

**Figure 17-1 Distance to GPI from cloud base (2 1/2° glide slope)**

## View of the Runway Environment

12.   If the distance to the Glidepath Ground Point of Interception (GPI) from where you break out of the cloud base is known, the view of the runway environment that should be able to be seen can be anticipated. This can be of considerable assistance when reaching the decision height. The distance is given for various cloud heights for a 2 1/2° glide slope in Figure 17-1 and a 3 glide slope in Figure 17-2.

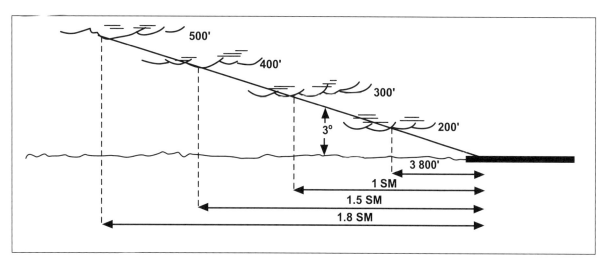

**Figure 17-2 Distance to GPI from cloud base (3° glide slope)**

13. Typical approach lights extend 3,900 feet from the GPI. The minimum visibility required to pick out the start of the approach lights from cloud base for 2 1/2° and 3° glide slopes is shown in Figure 17-3.

| Cloud Base | 2.5 Glide Slope | 3 Glide Slope |
|---|---|---|
| 500' | 7600'(1.55SM) | 5800' |
| 400' | 5200' | 3700' |
| 300' | 2900' | 1800' |
| 200' | 700' | — |

**Figure 17-3 Visibility required to detect the start of the approach lights**

14. With good visibility, the entire length of the runway will be visible from 200 feet or higher since the horizon at 200 feet is about 15 miles away. In considering what portion of the runway environment will be in view, the cockpit cut-off angle must be taken into consideration.

15. The position on the windscreen where the lights will first appear is dependent on the aircraft heading relative to the runway centre line and on the power setting, and these in turn are dependent on the wind. If the heading is to the right of the centre line, the runway and approach lights will appear left of their normal position on the windscreen and, if the heading is to the left of the centre line, they will appear to the right of their normal position. If more than usual power is required on the descent because of strong head winds, the lights will appear lower than normal, and if less power is required because of tail winds or light winds, the lights will be higher than normal.

**17** CHAPTER

## Slant Visual Range

16. The visibility out of the cockpit is the critical factor in being able to transit to visual flight. It depends on the slant visual range (SVR) plus the effects that precipitation, ice and so forth may have on the windscreen. There is no way of indicating what the visibility from the cockpit will be, however, if the available weather information is understood, a shrewd assessment can be made. The weather information that is relevant is prevailing visibility, runway visual range (RVR), type and intensity of precipitation and the various phenomena that reduce visibility such as fog or haze.

## Prevailing Visibility

17. "Prevailing Visibility" is the maximum visibility common to sectors comprising half or more of the horizon circle as viewed from the observing site at eye level. It is reported in statute miles and fractions of miles. A prominent object viewed against the horizon is used for day visibility, an unfocussed light of moderate intensity for night visibility. From the cockpit, you are viewing runway and approach lights and ground features and you are looking down on them obliquely from above. From the cockpit, the ground features on or near the runway tend to predominate during daylight and at night the lights predominate. There is a period of time at twilight where difficulty is encountered because ground features are fading out, yet it is not dark enough for lights to show up brightly.

18. If the prevailing visibility is less than 3 miles and it is fluctuating by one quarter or more, the range of visibility will be indicated in the remark group of the METAR. The fact that it is varying will generally imply that on your approach and landing run you will encounter patches of both reduced and improved visibility.

## Runway Visual Range

19. The runway visual range (RVR) is the maximum distance at which the runway, or the specified lights or markers delineating it, can be seen from a position above a specified point on its centre line. In other words, it is an instrumentally derived value that represents the horizontal distance a pilot may see down the runway.

20. The RVR is measured by a forward scatter meter located alongside the runway near its threshold. For CAT II landing systems, a second forward scatter meter is provided about two thirds of the way down the runway. The forward scatter meter is the instrument used to measure RVR. It is similar to the visibility instrument used by an AWOS. The sensor uses the forward scatter principle to measure the atmospheric extinction coefficient and extrapolates this into a visibility. Precipitation (snow, rain...) and obscuration to vision (fog, smoke...) affect this coefficient.

21. To calculate RVR, three factors must be known. The first is the visibility provided by the forward scatter meter, the second is the ambient light level and the third is the runway light intensity (controlled on request by the ATC controller). Those three factors are input into equations that are processed by computer. The result is the RVR.

22. The RVR is generally calculated every 15 seconds but can be calculated more frequently if the runway light intensity is changed. The ATC has a display that gives a one-minute average every minute. The weather office generally has a display that gives a 10-minute average every minute with a tendency.

The tendency indicates whether there was a distinct upward or downward change, or no distinct change in RVR values during that ten-minute period.

23. The readings are averaged over a period of time because during periods of low visibility there are often large fluctuations in visibility over extremely short time intervals. The average will give a more meaningful RVR since on either approach or take-off you are moving through a block of air that contains similar fluctuations of visibility.

24. An RVR is reported when the prevailing visibility is 1SM or less, or the RVR value for the designated runway(s) is 6,000 feet or less. The RVR is measured in increments of 100 feet up to 1,000 feet, increments of 200 feet from 1,000 feet to 3,000 feet, and increments of 500 feet above 3,000 feet to 6,000 feet. If there is a large variation of the RVR during the 10-minute average, the minimum one-minute average and maximum one-minute average during that 10 minutes will be given instead.

25. On some occasions the RVR will not accurately represent what the pilot sees. Those conditions generally occur during the day with shallow fog or with a snow surface. The glare caused by the sun reflecting on those surfaces will affect the pilot's view. In those situations, the use of prevailing visibility would be more appropriate.

26. You may use RVR instead of prevailing visibility for landing and take-off minima, but only for runways equipped with a forward scatter meter. In such cases the table in Figure 17-4 can be used.

| Ground Visibility | | RVR |
|---|---|---|
| 1 mile | | 5 000 feet |
| 3/4 mile | - EQUAL - | 4 000 feet |
| 1/2 mile | | 2 600 feet |
| 1/4 mile | | 1 200 feet |

**Figure 17-4 Equivalent visibility from RVR**

## Phenomena Reducing Cockpit Visibility

27. The more common phenomena aside from cloud that will reduce the visibility from the cockpit are fog, haze, smoke, snow, rain, drizzle, blowing snow and blowing dust.

28. The predominant types of fog are "Radiation," "Advection," and "Frontal." Radiation fog forms under nearly calm conditions at night with clear skies. It tends to form first as a patchy shallow layer near the ground with the tops slowly rising during the night. In Figures 17-5, 17-6 and 17-7, the slant visual range (SVR) as you descend from 200 feet down to the surface is indicated by the curve. For example, in Figure 17-5 the SVR at 200 feet is 3,500 feet and it decreases to a minimum of 1,500 feet near the top of the fog at 70 feet. These figures are examples only, the SVR will vary in specific situations. Figure 17-5 illustrates the situation while the fog is still shallow. Since the night is clear and the fog shallow, you will have visual contact with the runway as you start the approach. The visibility will begin to reduce rapidly as you pass your decision height, dropping from 3,500 feet at 200 feet to 1,500 feet at 70 feet. The clue that this might happen would have been the reported RVR and prevailing visibility and the fact that the sky condition would not be totally obscured. The depth of fog

17 CHAPTER

necessary to totally obscure the sky is about 300 feet. There might also have been a statement in the report indicating that the fog was thin or that the tower visibility was good.

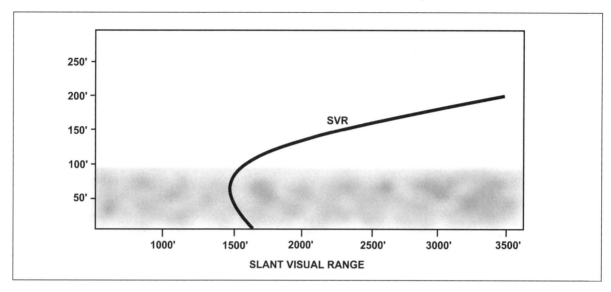

**Figure 17-5 Shallow fog**

29. As the fog becomes more mature, and the top rises, there is a tendency for the visibility in the lowest layer to improve slightly, but to decrease above this. Cockpit visibility will be fairly uniform and lower than the reported RVR until a fairly low level is reached when it will improve up to the RVR (Figure 17-6).

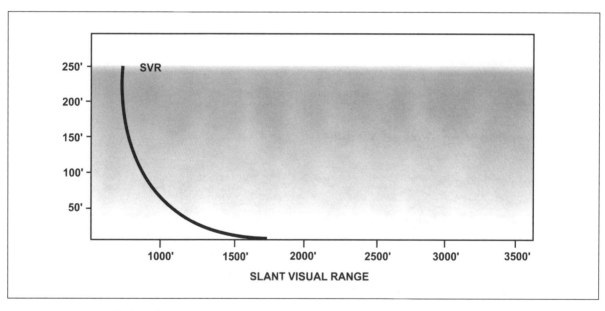

**Figure 17-6 Mature radiation fog**

30. Radiation fog will normally not dissipate until after sunrise. The initial stirring of the air near sunrise will tend to both reduce visibility throughout the layer and to increase the depth of the fog. With further heating, the fog will then become patchy, resulting in a variable RVR. The visibility will improve from the surface up and the fog will frequently lift into stratus fractus which will then dissipate.

31. Advection fog occurs when moist air moves over a colder surface. It is typical over oceans and coastal areas but can also occur inland. If the sun's rays can penetrate the fog over land there will be a tendency for a slight improvement near the surface. The SVR indicated in Figure 17-7 shows a steady improvement as the altitude decreases. The fog may get patchy, and this will be indicated by a varying RVR. On the other hand, if the sun's rays cannot penetrate the fog, because it is night or because of overlying cloud, or if there is an abundance of industrial pollution, the surface visibility will remain very low, and there will be little or no visibility improvement as at lower altitudes. The RVR will indicate the low visibility.

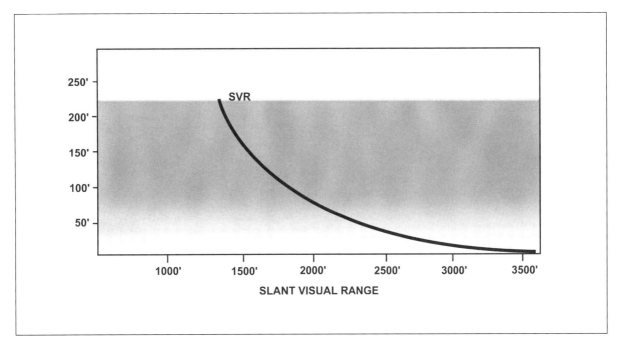

**Figure 17-7 Advection Fog**

32. Warm frontal fog forms as a result of moisture added to the air through the evaporation of relatively warm rain as it falls into colder air under the frontal surface. The SVR will generally be similar to that shown in Figure 17-7 with poor and uniform visibility for most of the approach, improving slightly in the very lowest levels.

33. SVR in snow at the decision height will generally be only slightly less than the reported RVR, so that there will be little improvement while in descent. The use of landing lights at night, however, may cause an overall brightness and reflection from the snow in front that will make it difficult to see the runway or approach lights and can cause disorientation.

34. The SVR in rain at the decision height will also be very similar to the RVR. The rain striking the windshield, however, will further deteriorate visibility. The extent of this will depend upon aircraft type. There is another effect of rain that should be made apparent. The rain on the windshield will cause a refraction so that the runway will appear slightly lower than it actually is. The cockpit visibility in rain can be assessed from the intensity of the precipitation being reported, from the prevailing visibility and from the RVR.

17 CHAPTER

35. The extent of blowing snow and blowing dust is dependent on the wind speed and on the instability of the air. They present an SVR situation similar to that of shallow fog as represented in Figure 17-5. The cockpit visibility may initially be good all the way down to decision height, and then deteriorate abruptly as the flare is initiated. An assessment of the situation can be made from the wind speed, the prevailing visibility, the RVR and whether or not there is an obscured sky condition being reported. A PIREP indicating the top of the blowing condition or a report that the tower visibility is good can be helpful. During blowing snow, blowing dust or drifting snow, you can expect a false drift sensation which can result in a landing with drift.

36. The RVR can be quite unrepresentative if the landing is towards a low sun and there is a restriction to visibility occurring. The cockpit visibility into fog, haze, ice crystals or smoke will be lower than either the prevailing visibility or the RVR under these conditions. In such situations, if the wind permits, the use of a different runway could be requested.

# Summary - Chapter 17

- Ceilometers measure the cloud base at the observing site.

- There will frequently be fragments of cloud obstructing your vision below the level measured as the cloud base.

- The cloud height provided, even at low levels, is rounded off to the nearest 100 feet.

- In obscured conditions, cockpit visibility will always be reduced.

- The amount of the runway environment that will be visible at your height when you break out of cloud can be assessed from the RVR.

- The position where the runway lights will appear on the windscreen can be assessed from the power setting and heading required because of the approach wind.

- Prevailing visibility is a human assessment of the general visibility surrounding the aerodrome.

- Runway Visual Range (RVR) is an assessment of the distance runway lights can be seen measured by a forward scatter meter near the runway threshold.

- Slant Visual Range (SVR) is the visibility from the cockpit.

- Radiation fog forms as a patchy shallow layer, thickening and intensifying throughout the night, especially at sunrise. It then frequently lifts to a stratus fractus layer before dissipating due to daytime heating with an improvement in RVR.

- RVR in advection fog may improve over the land during daytime if the sun can penetrate it, otherwise there will be no improvement.

- SVR in snow will show little change with altitude. Landing lights can glare and cause disorientation at night.

- SVR in rain depends on windshield characteristics. Rain can cause a refraction that results in the runway appearing slightly lower than it actually is.

- SVR in blowing snow and blowing dust decreases as you approach the runway and gives a false drift sensation.

- RVR is probably the most useful information provided for landings; however, in abnormal lighting conditions, it must be used with caution.

**17** CHAPTER

# Chapter

# 18

Because weather has such a pronounced effect, it should always be considered when planning flying operations.

# CHAPTER 18

# AVIATION CLIMATOLOGY

## Purpose of Aviation Climatology

1. Aviation climatology provides information on the prevailing conditions of the various atmospheric phenomena that affect aviation. The information could be required for a specific location, for an area, for a route or for all three. In order to provide valid climate information, the data should be based on about 30 years of records.

2. Planning for the inauguration of regular operations usually requires considerable advance preparation necessitating the fullest information concerning meteorological conditions to be expected. Knowledge is required not only of average conditions but also of the extremes, or more particularly, of the frequency with which conditions exceed (or fall short of) certain specified limits. In some cases, the extremes of adverse conditions, such as head winds along a route or fog at terminal airfields, are too severe to allow for an efficient operation. It is advantageous, therefore, to know beforehand the frequency and likely duration of conditions that can make operations impracticable. If the frequency is appreciable, an operator may decide to plan for an expected regularity of, say, 85%, meaning that cancellations are not expected to exceed 15% of all occasions over some particular period. Hence detailed climatological information is required in order for the potentialities of a regular operation to be estimated.

## Methods of Presentation

### Tables

3. Climatic information concerning almost every aspect of aviation is available and is provided in several ways. Figure 18-1 shows general information for Moose Jaw including temperature, precipitation, amount of sunshine and snow depth. Although not directly related to flying operations, this type of material can be used for planning airfield maintenance such as heating, drainage and snow removal. For instance, from the table, it can be seen that the greatest 24-hour snowfall has fallen at Moose Jaw in September (27.2 cm). This means that to ensure continued operations at Moose Jaw, the runway snow removal equipment should be ready for the winter season by September.

**18** CHAPTER

## MOOSE JAW A, SASK.

| | JAN | FEB | MAR | APRIL | MAY | JUNE | JULY | AUG | SEPT | OCT | NOV | DEC | YEAR | PERIOD |
|---|---|---|---|---|---|---|---|---|---|---|---|---|---|---|
| **TEMPERATURE (°C)** | | | | | | | | | | | | | | |
| Absolute Maximum | 10.6 | 12.2 | 23.2 | 32.8 | 37.2 | 36.1 | 38.9 | 41.7 | 36.1 | 31.1 | 21.7 | 11.1 | 41.7 | 1943-73 |
| Year | 1968 | 1954 | 1966 | 1949 | 1960 | 1956 | 1959 | 1949 | 1967 | 1943 | 1965 | 1954 | 1949 | |
| Mean Maximum | -10.0 | -5.7 | -0.9 | 11.0 | 18.2 | 22.9 | 27.1 | 26.4 | 19.6 | 13.4 | 1.9 | -5.4 | 9.9 | 1943-70 |
| Mean Temperature | -15.3 | -11.3 | -6.3 | 4.4 | 11.2 | 16.2 | 19.7 | 18.8 | 12.6 | 6.7 | -3.2 | -10.4 | 3.6 | 1943-70 |
| Mean Minimum | -20.8 | -16.9 | -11.7 | -2.2 | 4.2 | 9.4 | 12.3 | 11.2 | 5.6 | 0.0 | -8.4 | -15.4 | -2.7 | 1943-70 |
| Absolute Minimum | -45.6 | -36.1 | -38.9 | -25.0 | -11.1 | -2.8 | 1.1 | 0.0 | -12.8 | -21.1 | -31.1 | -40.0 | -45.6 | 1943-73 |
| Year | 1943 | 1962 | 1972 | 1954 | 1954 | 1969 | 1967 | 1957 | 1961 | 1957 | 1960 | 1961 | 1943 | |
| % Temperature ≤0°C | 92.0 | 90.1 | 72.4 | 28.7 | 3.2 | a | - | - | 2.9 | 19.5 | 71.4 | 85.8 | 38.7 | 1957-66 |
| % Temperature ≤-20°C | 33.3 | 21.6 | 6.6 | - | - | - | - | - | - | - | 4.3 | 19.6 | 7.1 | 1957-66 |
| % Temperature ≥25°C | - | - | - | 0.6 | 4.7 | 10.2 | 24.0 | 20.0 | 5.1 | 0.9 | - | - | 5.5 | 1957-66 |
| **PRECIPITATION** | | | | | | | | | | | | | | |
| Mean Precipitation(mm) | 18.2 | 14.0 | 17.0 | 25.2 | 42.0 | 68.8 | 53.8 | 42.4 | 35.8 | 19.0 | 18.6 | 20.6 | 375.4 | 1943-70 |
| Greatest 24 hour Precipitation (mm) | 13.4 | 12.2 | 15.2 | 38.6 | 33.6 | 80.6 | 39.6 | 35.8 | 52.8 | 26.6 | 12.0 | 13.0 | 80.6 | 1943-73 |
| Greatest Rainfall (mm) | 2.2 | 4.0 | 10.0 | 59.4 | 79.2 | 162.8 | 127.8 | 158.2 | 103.6 | 67.6 | 7.2 | 4.4 | 500.2 | 1943-73 |
| Year | 1962 | | 1967 | 1955 | 1958 | 1944 | 1969 | 1954 | 1954 | 1969 | 1967 | 1956 | 1954 | |
| Mean Rainfall (mm) | 0.2 | 0.6 | 1.2 | 13.8 | 39.8 | 68.8 | 53.8 | 42.4 | 34.6 | 12.8 | 1.8 | 0.6 | 270.2 | 1943-70 |
| Least Rainfall (mm) | 0.0 | 0.0 | 0.0 | T | 2.8 | 12.0 | 4.0 | 0.8 | 1.2 | 0.6 | 0.0 | 0.0 | 177.8 | 1943-73 |
| Year | | | | | 1958 | 1961 | 1944 | 1961 | 1944 | 1970 | | | 1958 | |
| Greatest 24 hour Rainfall (mm) | 2.2 | 3.8 | 8.8 | 38.6 | 33.6 | 80.6 | 39.6 | 35.8 | 52.8 | 26.6 | 4.4 | 3.6 | 80.6 | 1943-73 |
| Greatest Snowfall (cm) | 46.2 | 35.8 | 47.6 | 68.4 | 7.6 | - | - | - | 37.6 | 23.6 | 49.6 | 60.4 | 259.0 | 1943-70 |
| Year | 1965 | 1972 | 1967 | 1970 | 1965 | | | | 1972 | 1969 | 1958 | 1973 | 1970 | |
| Mean Snowfall (cm) | 22.2 | 16.2 | 18.0 | 12.8 | 2.0 | - | - | - | 1.2 | 6.8 | 19.6 | 24.4 | 123.2 | 1943-73 |
| Least Showfall (cm) | T | 3.6 | 2.0 | 0.0 | 0.0 | - | - | - | 0.0 | 0.0 | T | T | 58.0 | |
| Year | 1944 | 1954 | 1960 | 1949 | | | | | | | 1953 | 1943 | 1968 | |
| Greatest 24 hour Showfall (cm) | 16.8 | 12.2 | 16.2 | 24.2 | 6.6 | - | - | - | 27.2 | 18.2 | 14.4 | 13.0 | 27.2 | 1943-73 |
| **SNOW DEPTH** | | | | | | | | | | | | | | |
| Maximum Snow Depth on Ground at Month End (cm) | 48 | 38 | 25 | 13 | - | | | | - | 8 | 20 | 30 | - | 1953-72 |
| Average Snow Depth on Ground at Month End (cm) | 13 | 10 | 3 | 3 | - | | | | - | - | 5 | 10 | - | 1953-72 |
| **SUNSHINE** | | | | | | | | | | | | | | |
| Possible No. of hours | 268 | 294 | 369 | 413 | 478 | 489 | 492 | 447 | 379 | 333 | 272 | 253 | 4487 | 1954-73 |
| Maximum | 137 | 283 | 235 | 310 | 333 | 382 | 400 | 368 | 268 | 242 | 167 | 147 | 2597 | 1954-70 |
| Mean | 100 | 173 | 160 | 214 | 267 | 274 | 344 | 296 | 207 | 175 | 101 | 83 | 2338 | 1954-73 |
| Minimum | 42 | 54 | 93 | 140 | 148 | 210 | 216 | 166 | 136 | 89 | 53 | 44 | 2007 | |
| **CLOUD COVER** | | | | | | | | | | | | | | |
| Mean Cloudiness (teths) | 5.4 | 5.2 | 5.2 | 5.2 | 5.1 | 5.2 | 4.3 | 4.1 | 4.8 | 4.8 | 5.8 | 5.4 | - | 1954-72 |
| % Freq. clear Skies | 17.6 | 18.9 | 18.1 | 16.8 | 13.7 | 9.2 | 13.6 | 20.3 | 20.3 | 20.9 | 15.4 | 18.2 | - | |
| % Freq. 1.10-5/10 | 32.4 | 33.8 | 35.5 | 35.7 | 41.3 | 43.5 | 52.0 | 47.0 | 37.0 | 36.8 | 30.8 | 31.7 | - | |
| %Freq;6/10-9/10 | 21.7 | 22.1 | 22.1 | 24.3 | 26.6 | 32.0 | 26.5 | 23.1 | 25.2 | 21.2 | 20.6 | 21.3 | - | |
| % Freq. Overcast | 28.3 | 25.2 | 24.0 | 23.2 | 18.4 | 15.3 | 7.9 | 9.6 | 17.5 | 21.1 | 33.2 | 28.8 | - | |

**Figure 18-1 Climatic data for Moose Jaw**

CHAPTER **18**

## Charts

4.   Another way of presenting climatic information is the chart as shown in Figure 18-2. This figure indicates the percentage of time that ceilings are less than 1,000 feet and/or visibilities are less than three miles over North America for each season of the year. Charts showing other limits are available. From these charts, it can be seen that the choice of the Canadian Prairies for flying training school locations is a good one since even in the worst season (winter) the weather is below 1,000 feet and/or 3 miles little more than 10% of the time. The largest percentage of time below 1,000 feet and/or 3 miles occurs in Canada in spring and summer over the eastern tip of Newfoundland.

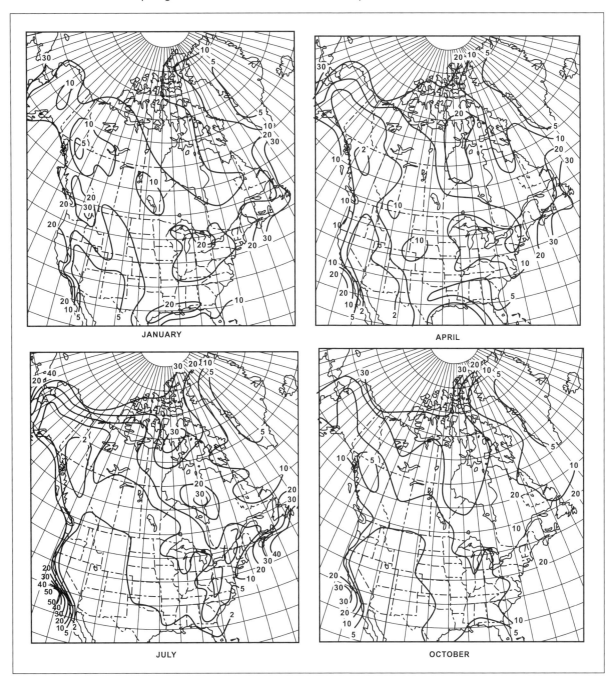

**Figure 18-2 Percent frequency of ceiling less than 1,000 feet and/or visibility less than 3 miles**

**18** CHAPTER

## Graphs

5. Figures 18-3 and 18-4 provide detailed and useful information on terminal conditions. They indicate the percentage of time that the weather is below 1,000 feet and/or 3 miles for each hour of the day and each month of the year. The time of sunrise and sunset is marked by a broken line on these charts. Charts based on other limits are also available. Figure 18-3 is for Cold Lake and Figure 18-4 for Shearwater and they illustrate the marked difference between continental and maritime terminals.

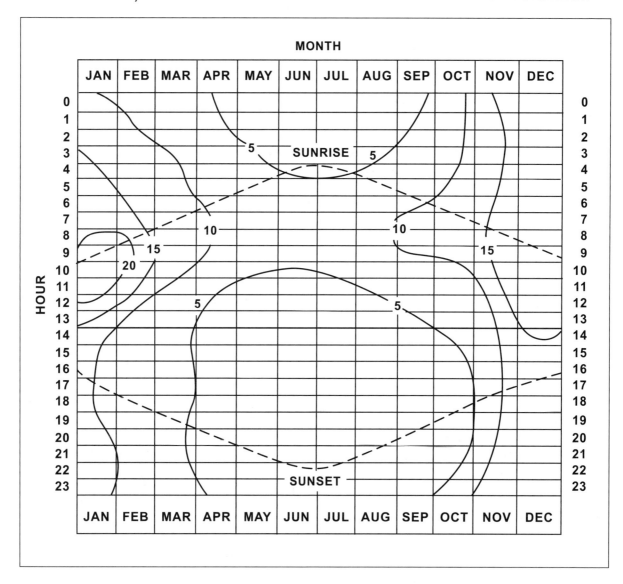

**Figure 18-3 Percent frequency of ceiling less than 1,000 feet and/or visibility less than 3 miles (Cold Lake)**

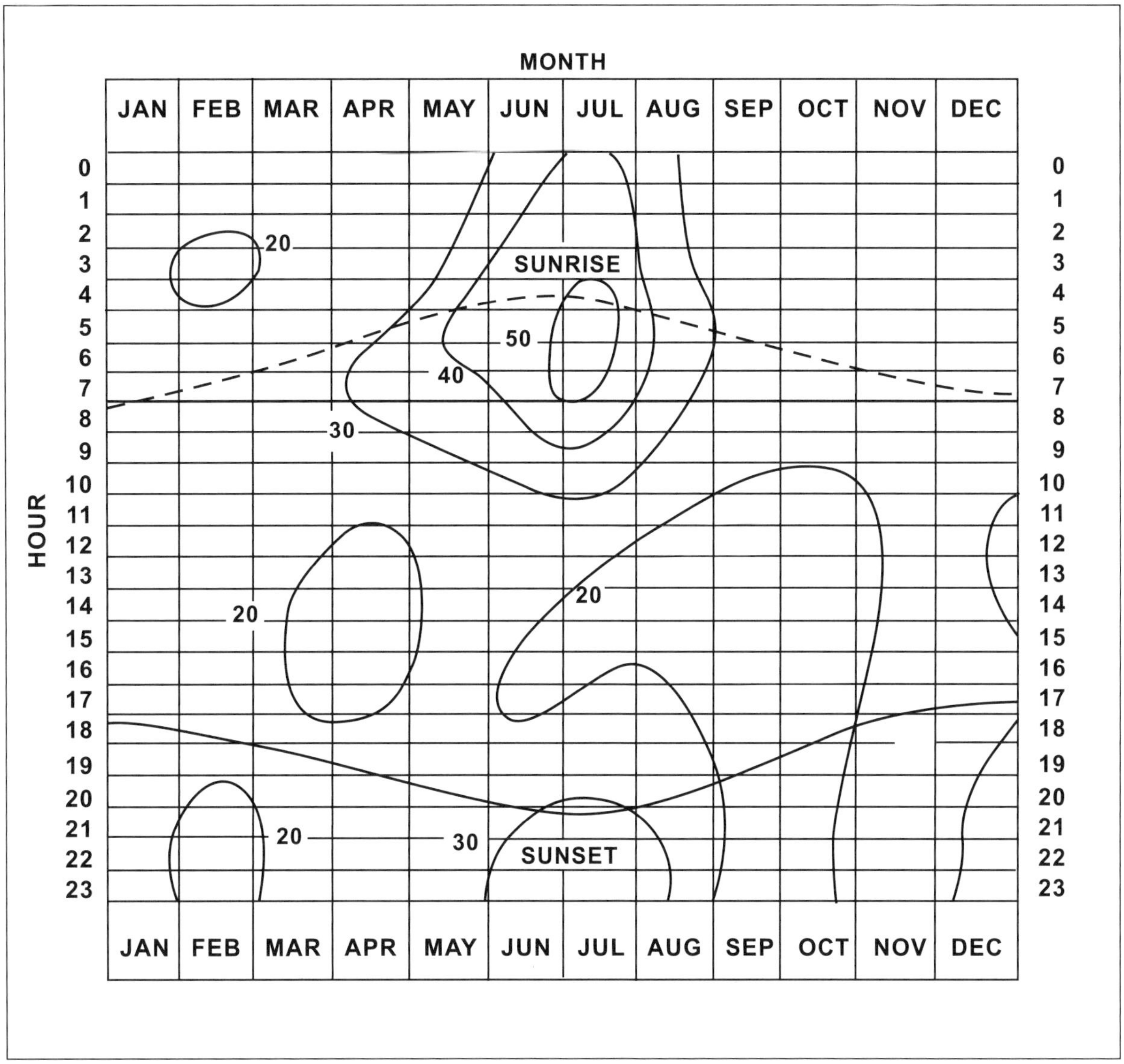

MONTH

| | JAN | FEB | MAR | APR | MAY | JUN | JUL | AUG | SEP | OCT | NOV | DEC |

**Figure 18-4 Percent frequency of ceiling less than 1,000 feet and/or visibility less than 3 miles (Shearwater)**

6. At Cold Lake, conditions are below 1,000 feet and/or 3 miles most frequently (a little over 20% of the time) just near sunrise in January. At Shearwater, they are below these limits over 50% of the time near sunrise in July. Charts such as these are useful for planning exercises, especially if the exercises are going to be repeated on an annual basis. Actual conditions in any one year may vary from the average but the odds are improved by using the climatic information.

7. An enormous amount of aviation climatological information has been published and is readily available. The above figures have provided only a few examples. Information that is not already available can be specially processed on request to the Canadian Forces Weather and Oceanographic Service.

## Summary - Chapter 18

- Climatic information is available for routes, areas and terminals.

- About 30 years of records are required for valid climatic information.

- This information provides assistance for planning all types of flying operations.

- Information not already available can be obtained by special request.

# Chapter
# 19

Each area of Canada has its own weather peculiarities and hazards. To fly safely across the country, you must know what they are.

# CHAPTER 19

# WEATHER ACROSS CANADA

## Geographic Features

1.  A major influence on the type of flying weather that occurs over oceans and along coastlines is the water temperature. This temperature is controlled primarily by ocean currents. The North Pacific Current supplies relatively warm water along the West Coast of Canada. This warm water current maintains the Gulf of Alaska ice-free through the winter.

2.  Along the East Coast, the Gulf Stream flows out of the hot Caribbean and the Gulf of Mexico. The Labrador Current, cold water complete with icebergs, flows from the Baffin Bay area along the coasts of Labrador, Newfoundland and Nova Scotia.

3.  On land, the Rocky Mountains and the Coastal Range barrier extend unbroken from Alaska to Mexico with peaks extending up to 10,000 to 15,000 feet. To the east of this mountain barrier, the Great Central Plains extend low and flat, unbroken from the Gulf of Mexico to the Arctic Ocean. Northern Canada is covered with lakes that are open and providing moisture to the air in the summer, but frozen in the winter. The Appalachian Mountains lie to the east with peaks extending to 4,000 and 5,000 feet.

4.  The permanently ice-capped Arctic Ocean lies to the north of Canada. To the south of the United States are the vast areas of warm Pacific and Atlantic Ocean waters whose temperatures remain between 20°C and 30°C year-round.

## Pressure Systems

5.  The dominant features are the semi-permanent high-pressure areas of low latitudes and the semi-permanent low-pressure areas of high latitudes. Both of these systems vary greatly in intensity from day to day and they also follow a definite seasonal trend.

6.  A semi-permanent "Pacific High" extends from the West Coast of the United States to west of Hawaii. Its Atlantic counterpart is the "Bermuda High," which dominates the area between Bermuda and the Azores. Both of these high-pressure areas are more extensive and have higher pressures in summer than in winter.

7.  A semi-permanent low-pressure area called the "Aleutian Low" occurs in the Gulf of Alaska and over the Aleutian Islands. It has an Atlantic counterpart called the "Icelandic Low," which frequently lies in the region between Greenland and Iceland. These semi-permanent low-pressure areas are very much more intense in winter than in summer.

**19** CHAPTER

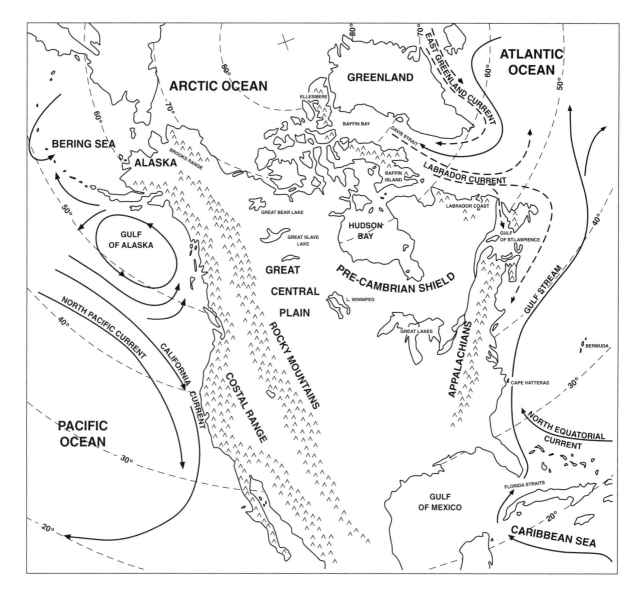

**Figure 19-1 Geographic features controlling the weather in Canada**

8.   Figures 19-2 and 19-3 illustrate the semi-permanent pressure systems during the months of January and July. These charts are based on pressure readings averaged over the entire month. Charts for a particular time could be quite different and the pressure centres could be much more intense.

9.   Migratory frontal depressions occur on the Polar, Maritime and Arctic frontal systems. These are also much more intense in winter than in summer. Their tracks are variable and they move with the general airflow within the troposphere. Frequently, these lows will intensify if they curve to the left and move towards the north.

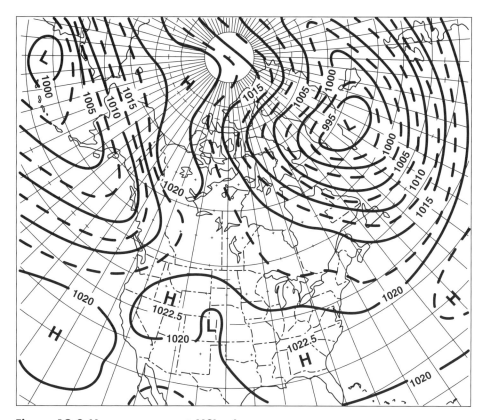

**Figure 19-2 Mean pressure at MSL - January**

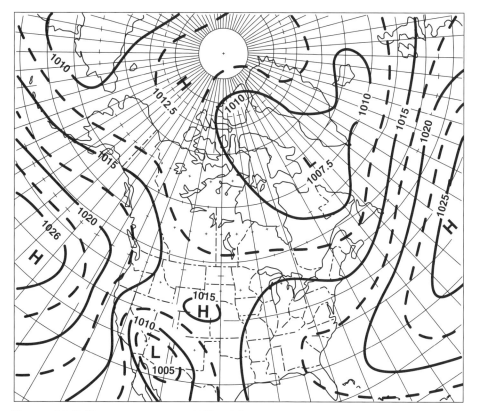

**Figure 19-3 Mean pressure at MSL - July**

10. The semi-permanent and migratory pressure systems together with the various geographical features that have been described produce the flying weather in which we are interested. Figures 19-4 and 19-5 provide an overview of the general flying weather in the summer and the winter seasons by indicating the percentage of times that conditions are below ceilings of 1,000 feet or visibilities less than 3 miles over North America. Because of the small scale of the charts, the percentage lines drawn on them are smoothed. Particularly in coastal and mountainous regions there are small pockets where the weather varies from that indicated by the percentage lines. Conditions at terminals within these pockets may vary somewhat from the percentages indicated on the charts.

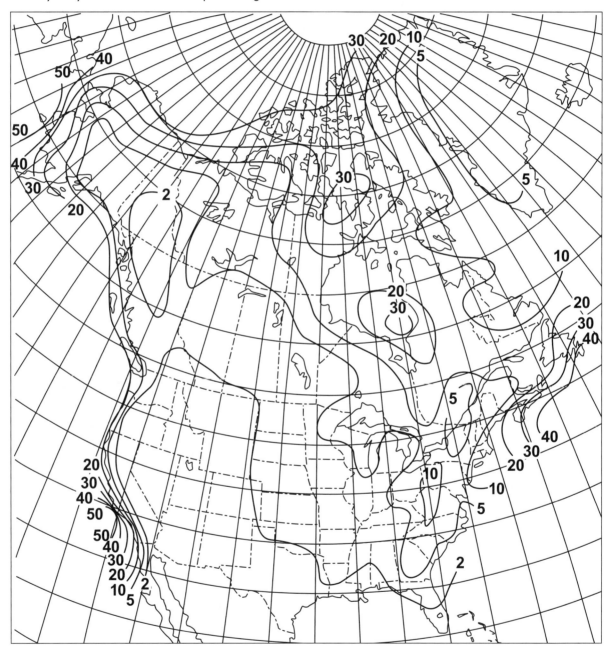

**Figure 19-4 Percent frequency of ceilings less than 1,000 feet and/or visibilities less than 3 miles - July**

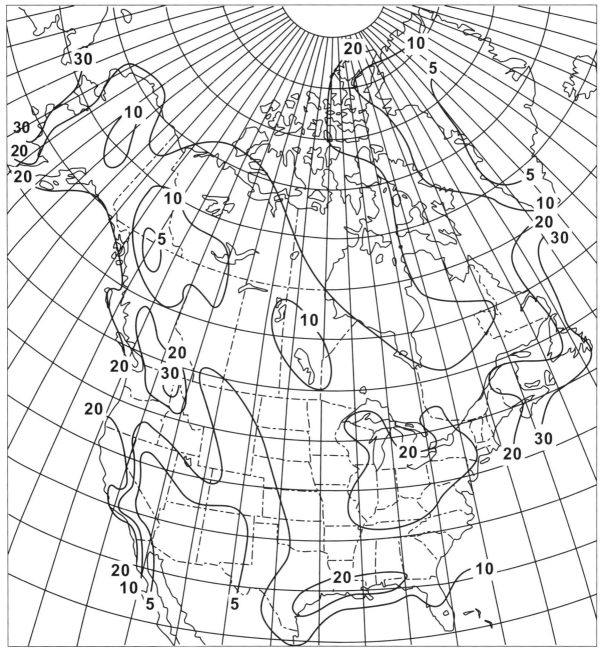

**Figure 19-5 Percent frequency of ceilings less than 1,000 feet and/or visibilities less than 3 miles - January**

## West Coast Weather

11. The West Coast of Canada is generally exposed to a westerly flow of moist air from the Pacific. Due to the warm North Pacific Current this westerly flow is usually mild so that even in winter below-freezing temperatures seldom occur. The mountains normally act as a barrier to prevent Continental Arctic air from the north or east from advancing westward to the coast. Occasionally, however, the cold air builds up east of the mountains to such an extent that it spills over onto the coast through valleys and inlets. As well as abnormally cold temperatures this produces strong boundary layer winds and wind shears that are hazardous to flying.

12. During winter, intense low-pressure systems develop over the Pacific and move on to the West Coast. These storms combined with the orographic lift over the mountains produce extensive and thick cloud shields with heavy precipitation. Rain usually occurs at lower elevations and snow occurs on the mountain tops. Low ceilings and visibilities and heavy icing occur. Surface winds will usually be strong with local topographical effects producing unexpected directions.

13. In the wake of these winter storms, the West Coast frequently comes under the influence of a flow of air that originates as Continental Arctic air over Alaska or Siberia. After flowing across the Gulf of Alaska, this air arrives over the West Coast as very moist and unstable Maritime Arctic air. This gives rise to extensive convective cloud, producing frequent heavy showers of rain or wet snow and very turbulent flying conditions in and below the cloud.

14. During the summer the Pacific lows are very much weaker and the Pacific semi-permanent high-pressure system generally dominates the West Coast weather.

15. Figure 19-6 shows the marked variation in flying weather between the different seasons of the year by indicating the percentage of time ceilings are less than 1,000 feet and/or visibilities are less than 3 miles for each month of the year and each hour of the day at Comox and Vancouver. The period of time just after sunrise throughout fall and winter is particularly bad at Vancouver.

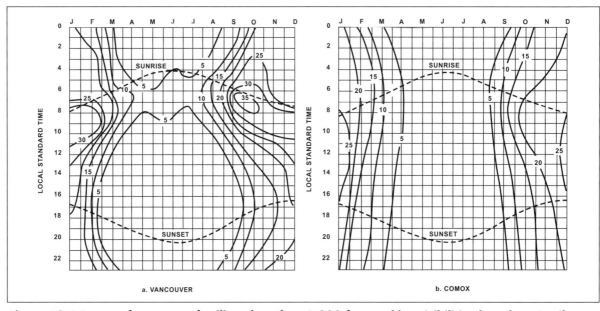

**Figure 19-6 Percent frequency of ceilings less than 1,000 feet and/or visibilities less than 3 miles**

## Mountain Weather

16. The eastern slopes of the Coastal and Rocky Mountain ranges have less cloud and precipitation than the western slopes. This is because the air flows predominately from the west and undergoes orographic lift on the western slopes. Low ceilings and visibilities are much more predominant in winter than in summer.

17. During the summer frequent afternoon thunderstorms develop due to combined effects of daytime heating and orographic lifting. Still, summer rainfalls tend to be light.

18. In the mountains, summer temperatures tend to be higher than West Coast temperatures, except at high elevations. Winter temperatures tend to be lower than those on the coast, and can be very low due to invasions of cA air combined with radiational cooling. Some valleys have a high incidence of fog during winter months set off by radiational cooling and cold air drainage.

19. Jet streams from the west or southwest will frequently set off mountain waves. These will be stronger and more dangerous during the winter due to stronger jet stream winds. It is mountain waves that produce chinook winds on the Prairies.

20. Anabatic, katabatic, funnel and valley winds are all features of the mountain area and can cause extreme turbulence and shears.

## Great Central Plains Weather

21. Stretching from the Arctic to the Gulf of Mexico, the Great Central Plains is an area of light precipitation, little cloud and usually good flying weather. Because there are no east-west mountain barriers across the plain there is an easy path for cold air from the Arctic to surge far southward and for warm air from the south to penetrate into northern latitudes. The western mountains are a pronounced barrier to the movement of air onto the plains from the Pacific.

22. Very large temperature variations occur over the plains. During the winter, Continental Arctic air, fresh out of the Arctic, can produce extreme cold. Conversely, during the summer, Maritime Tropical air from the Gulf of Mexico can reach very high temperatures with daytime heating. Precipitation, although small in amount, is at a maximum in summer and occurs principally due to convection. Low ceilings and visibilities are infrequent but occur most often during the winter (Figures 19-4 and 19-5).

23. Frontal depressions tend to form just east of the Rockies during the winter. Those forming in Alberta frequently move rapidly east-southeast bringing a surge of cold air and occasionally blizzard conditions behind them. The storms that become the most intense and produce the most weather develop in the central United States just east of the Rockies and move north or northeast into Canada. Maritime Tropical air from the Gulf of Mexico may be injected into these storms providing enough moisture to produce severe weather and heavy precipitation (rain or snow) from southern Saskatchewan to the Gulf of St. Lawrence.

24. Because the Great Plain slopes upward to the Rocky Mountains, an east or northeast wind is upslope. This flow will occur to the north of a low-pressure system. If the air is moist, an extensive area of low cloud or fog may occur. Radiation fog occurs very seldom because the air is normally too dry within the high-pressure systems that are required for the fog to form.

**19** CHAPTER

25. An outbreak of Continental Arctic air during the fall can cause very heavy snow showers over and in the lee of open lakes. This can cause a sharply delineated area of very low weather that will persist until the airflow changes direction. The term "Lake Effect" may be used to describe the situation.

26. Across the Great Plains precipitation and clouds during the winter occur mainly due to low-pressure systems with frontal activity. Summer weather is mainly of the convective type due to unstable air having its instability released by frontal action.

## Great Lakes Weather

27. The large water surface of the Great Lakes has an effect on the weather occurring over and around them. Generally there is more cloud and precipitation over the Great Lakes than the Great Central Plains. Also winter temperatures are not as cold, and summer temperatures not as hot as those of the Great Central Plains. Precipitation is very uniformly spread throughout the year and does not show the summer maximum of the Prairies, although low weather is slightly more prevalent in the winter.

28. Most of the Great Lakes are large enough and deep enough that they do not freeze over during the winter. When outbreaks of Continental Arctic air flow across this open water in late fall or winter, "Lake Effect" develops. With the general west or northwest flow, this produces cloudy skies and increased snowfall with low ceilings and visibilities to the south and east of the lakes. The cumulus clouds that develop in the lake effect usually are topped at 5,000 to 10,000 feet. Aircraft icing in these clouds can be a serious problem and ceilings and visibilities vary widely and rapidly.

29. During the spring, if Maritime Tropical air flows over the relatively cold water of the Great Lakes, advection fog or low stratus cloud may develop. During the summer months with a flow of Maritime Tropical air over the Great Lakes, the cooling effect of the water will reduce the instability of the air, thus producing less violent thunderstorms and fewer tornadoes on the lee sides of the lakes than elsewhere.

30. During the summer, sea breezes develop along the shores of the Great Lakes and this reduces the maximum temperatures.

## East Coast Weather

31. East Coast temperatures are not moderated by the nearby ocean to the extent that the West Coast temperatures are because the general flow is westerly, from the interior of North America. Thus, East Coast winter temperatures are colder and summer temperatures are hotter than those on the West Coast.

32. The East Coast is notorious for its winter depressions that intensify rapidly and produce widespread clouds, heavy precipitation and strong winds. The condition that favours this development is a large temperature gradient (frontal zone) along the East Coast. This results when very cold air covers the land and very warm and moist air lies over the warm water of the Gulf Stream. The usual area for frontal depression to form is near Cape Hatteras, North Carolina.

33. These storms tend to track northeastward towards Newfoundland along the temperature change zone between the warm water of the Gulf Stream and the cold water of the Labrador Current (Figure 19-1). The Maritime Tropical air from over the Gulf Stream injects heat energy and moisture into the storm and this contributes to the development of clouds, precipitation and strong winds. Another contributing factor is the orographic lift that occurs as the east and northeast winds ahead of the low flow off the ocean, across the coastal plain, and up against the Appalachians. Heavy rainfalls, paralyzing snowfalls and hurricane force winds may occur with these storms.

34. These storms may also track through the Bay of Fundy, across New Brunswick and the Gulf of St. Lawrence then into Quebec before turning eastward out into the Atlantic. This storm track gives cloud and precipitation to Labrador and as far west as Eastern Ontario and James Bay.

35. In the wake of these intense East Coast storms there usually occurs an outbreak of fresh cA air in the form of strong northwesterly winds. Lake effect produces extensive stratocumulus and low-level cumulus clouds with showers in the lee of the Great Lakes, over the Appalachian Mountains and out over the Atlantic. Due to the strong winds and the heating from the Great Lakes, vigorous mechanical turbulence develops over and in the lee of the Appalachians. Icing may be a significant problem in these clouds. The clouds tend to dissipate in the subsidence in the lee of the Appalachians, but the turbulence may persist, and perhaps be augmented by mountain waves.

36. A major problem along the East Coast is the advection fog that develops, particularly during the summer, when Maritime Tropical air from over the Gulf Stream is advected over the cold waters of the Labrador Current. This frequently causes fog in the coastal areas from Cape Cod northward to Nova Scotia and Newfoundland. As long as the wind blows from the south or southeast the fog will persist. The semi-permanent high-pressure system over Bermuda may keep this flow up for days.

37. Frontal depressions are much weaker and less numerous in summer than in winter. Convective activity along cold fronts in summer can cause some problems particularly on the west side of the Appalachians where orographic lift intensifies the frontal systems.

38. Figures 19-4 and 19-5 indicate that the largest percentage of time that ceilings are below 1,000 feet or visibilities are below 3 miles occurs in the summer along the coastal areas, but that inland there is not a great deal of difference between winter and summer.

39. Figure 19-7 illustrates the marked variation that can occur within short distances in coastal areas due to terrain features. This figure indicates the percentage of time that conditions are below 1,000 feet and/or 3 miles for each month of the year and each hour of the day at Shearwater and Greenwood.

40. Shearwater is on the coast and Greenwood is inland in a valley. In both cases low weather conditions occur most frequently in the summer during the morning hours just after sunrise. The weather is below 1,000 feet and/or 3 miles over 50% of the time during this period at Shearwater and over 30% of the time at Greenwood. This is in large part due to advection fog.

41. Daytime heating causes an improvement during the afternoon at both terminals. At Shearwater, the improvement is to around 20% below 1,000 feet and/or 3 miles, and at Greenwood a marked improvement to around 4%. This improvement does not occur over the open ocean.

**19** CHAPTER

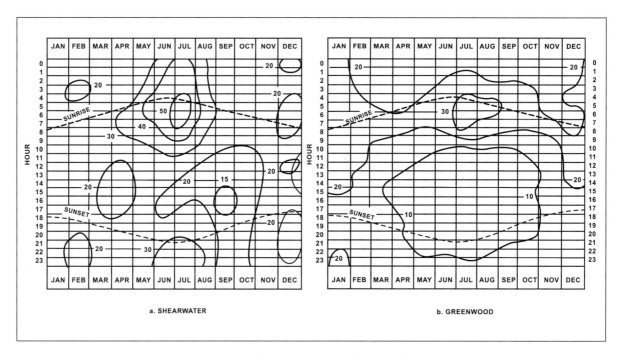

**Figure 19-7 Percent frequency of ceilings less than 1,000 feet and/or visibilities less than 3 miles**

## Hudson Bay Weather

42. Unlike the Great Lakes, Hudson Bay freezes over every year, usually by the end of December. Break-up of the ice starts during June, and the Bay is usually ice-free by the end of August. Refreezing will commence during October.

43. After freeze-up, open leads (patches of open water) will appear from time to time due to the action of winds and tides. Over and in the lee of the open leads, patches of ice fog will appear, but no other significant weather occurs.

44. During the summer, the cold water and ice surface cause advection fog to form when warm moist air drifts over the Bay. Once formed, this fog may move on shore and affect the coastal areas, producing low ceilings and visibilities.

45. The most significant and widespread weather that the Bay develops is that produced during the late summer and fall when outbreaks of arctic air sweep across its open waters. These outbreaks usually occur with northwesterly flows so the land areas affected are south and east of the water. Rain and snow showers develop in the lee of the Bay over northwestern Quebec and Northern Ontario. Hills may contribute orographic lift to the flow, resulting in low ceilings and visibilities that fluctuate widely. Moderate to heavy icing will occur in the cloud. As the freeze-up progresses these conditions will be reduced and will eventually be eliminated when the freeze-up is complete (Figure 19-4).

## Arctic Weather

46. The Rocky Mountains in Alaska, Yukon and British Columbia tend to block Pacific storms and air masses from entering the Arctic. To the east of these mountains the land is low and flat, including the western Arctic islands, which are really just a northward extension of the Great Central Plains. In the eastern Arctic, mountain peaks rise to 5,000 feet and 8,000 feet on Ellesmere, Devon and Baffin islands. Further east, coastal mountains in Greenland extend to over 9,000 feet.

47. North of the Arctic Circle there are three months of darkness during the winter. Terrestrial radiation during this period produces very cold surface temperatures and a surface-based inversion. A high-pressure area develops in the western Arctic and a trough develops over the eastern Arctic that extends out of the Icelandic low. The cold temperatures cause altimeters to seriously overread.

48. The period of maximum ice and snow cover occurs near the end of March or the beginning of April and this coincides with the time of best flying weather. A continental type of climate is prevalent in the winter. The main problem is keeping personnel and machines operating in the frigid temperatures and wind chills.

49. Winter storms occur infrequently. Occasionally, in the winter, a low from an occluded frontal wave will move into the Arctic through Bering Strait or Davis Strait. The lows will carry only sufficient moisture with them to produce layers of diffuse ice crystal cloud and light snow. The pressure gradients, however, may be steep, and the resulting strong winds can whip the fresh snowfall into blowing snow conditions with very poor surface visibilities. Because the low is associated with an occluded wave it becomes almost stationary, so these conditions can last for days.

50. Visibility can be reduced in other ways. In late winter or early spring, "Arctic haze" may reduce horizontal and slant visibilities. It may be encountered as high as 30,000 feet. Another winter problem is ice fog. At temperatures below -30°C, moisture from combustion or any other activity that produces moisture will saturate the air and cause ice fog to form.

51. Particularly in fall and winter, the topography has a marked effect on the boundary layer winds. Barrier, funnel and katabatic effects can be extreme and produce hazardous winds and wind shears.

52. Summer in the Arctic is a three-month period of continuous daylight. Surface temperatures rise to generally just above freezing over the water and along the coasts and to around 15°C to 20°C inland. As the temperature rises, open water appears and the climate becomes basically maritime in character. The high-pressure area and trough decrease in intensity and the pressure gradients become very slack.

53. With open water as a source of moisture, cloud becomes prevalent. Minimum ice cover occurs around late August or early September and this is the period of lowest flying conditions. The cloud is predominantly stratus, although cumulous and even cumulonimbus may form inland on the larger islands.

54. Warm air from the south occasionally penetrates to the Arctic during the summer causing advection fog to form over the water and along the coasts.

**19** CHAPTER

# Summary - Chapter 19

- **West Coast**

- Strong valley and inlet outflow winds.

- Intense winter lows.

- Convective turbulence over water and windward slopes.

- Lowest weather in fall and winter.

- **Mountains**

- Cloud most prevalent on windward slopes.

- Fog frequent in valleys in winter.

- Mountain waves.

- Anabatic, katabatic, funnel and valley winds.

- Lowest weather in winter.

- **Great Central Plain**

- Temperature extremes.

- Frontal depressions forming in Alberta and moving east or southeast may be followed by blizzards.

- Frontal depressions forming in the United States east of the Rockies and moving northeast produce the worst weather.

- Upslope stratus and fog can blanket the Prairies when a low lies to the south.

- Lake Effect occurs over open lakes in the fall.

- Low weather is infrequent but occurs mainly in the winter.

- **Great Lakes**

- Do not freeze over in the winter.

- Lake Effect in fall and winter.

- Advection fog in spring.

- Worst storms occur in winter when depressions move up from the west central states.

- Low weather is slightly more prevalent in the winter.

- **East Coast**

- Severe winter storms originate in Cape Hatteras, North Carolina area.

- Advection fog occurs in summer in coastal areas and over the open water.

- Lowest weather occurs in the summer in coastal areas but there is little variation elsewhere.

- **Hudson Bay**

- Freezes over in the winter.

- Ice fog occurs over and in the lee of open leads in winter.

- Advection fog occurs over and in the lee of the Bay in summer.

- Worst weather occurs in fall when cold air flows over open water.

- **Arctic**

- Best flying weather is in late winter or early spring.

- There are few winter storms, but blowing snow can occur when they happen.

- Ice fog occurs near inhabited areas in winter.

- Barrier, funnel and katabatic winds are prevalent in mountainous areas in winter.

- Advection fog occasionally forms in summer.

- Lowest weather occurs in late summer and early fall.

**19** CHAPTER

# AFTERWORD

1. In the Preface of this manual, it was stated that through the practical application of learned material you can develop the skill and judgement essential to your success as an aviator. Now at the conclusion, this statement must again be made.

2. Various features of the atmosphere that are important to aviation have been described. The atmosphere is the environment in which you operate and examples of the material that has been presented here constantly surround you. Look around and you will see them.

3. Aviation meteorology should be thought of as a part of airmanship and not as an isolated topic separate from flying. Weather will affect you. Your ability to cope with it will depend on your judgement. Your judgement, in turn, will depend upon your knowledge and experience. Experience can be gained rapidly or slowly depending upon the effort you make to recognize and understand the effects and influence of the atmosphere as you are flying. This manual has provided you with basic information. It is now up to you.

## GOOD FLYING

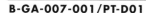